MW00680083

Agriculture in Urban Planning

Agriculture in Urban Planning

Generating Livelihoods and Food Security

Edited by Mark Redwood

International Development Research Centre
Ottawa • Cairo • Dakar • Montevideo • Nairobi • New Delhi • Singapore

publishing for a sustainable future
London • Sterling, VA

First published in the UK and USA in 2009
by Earthscan and the International Development Research Centre (IDRC)

Copyright © International Development Research Centre 2009

All rights reserved

ISBN: 978-1-84407-668-0

IDRC publishes an e-book edition (ISBN: 978-1-55250-427-7)
For further information:
International Development Research Centre
PO Box 8500
Ottawa, ON K1G 3H9, Canada
info@idrc.ca/www.idrc.ca

Typeset by 4word Ltd, Bristol
Printed and bound in the UK by MPG Books Ltd, Bodmin, Cornwall
Cover design by Yvonne Booth

For a full list of publications, please contact:

Earthscan
Dunstan House
14a St Cross St
London EC1N 8XA, UK
Tel: +44 (0)20 7841 1930
Fax: +44 (0)20 7242 1474
Email: earthinfo@earthscan.co.uk
Web: **www.earthscan.co.uk**
22883 Quicksilver Drive, Sterling, VA 20166-2012, USA

Earthscan publishes in association with the International Institute for Environment
and Development

A catalogue record for this book is available from the British Library

Library of Congress Cataloging-in-Publication Data

Agriculture in urban planning: generating livelihoods and food security
edited by Mark Redwood
    p. cm.
Includes bibliographical references.
ISBN 978-1-84407-668-0 (hardback)
1. Urban agriculture–Developing countries. 2. Sustainable development–Developing
countries. I. Redwood, Mark.
S494.5U72A42 2009
630.9173'2–dc22

                                                                          2008023954

The paper used for this book is FSC-certified and totally
chlorine-free. FSC (the Forest Stewardship Council) is
an international network to promote responsible management
of the world's forests.

Mixed Sources
Product group from well-mana
forests and other controlled s
www.fsc.org Cert no. SA-CO
© 1996 Forest Stewardship Coun

Contents

List of Figures and Tables

FIGURES

TABLES

Foreword

In 1998, the AGROPOLIS programme was launched by the International Development Research Centre (IDRC) to encourage young scholars to undertake research on urban agriculture (UA). In seven years, the programme granted more than 60 awards to graduate students from places as diverse as Argentina, the Congo and Canada, which were worth approximately US$1.2 million.

AGROPOLIS awardees have helped build the new field of UA and launch it into the mainstream as a viable tool for fighting hunger and poverty in cities. Their research projects have helped to generate new thinking in the field for the betterment of the poor. AGROPOLIS awardees have gone on to teach and continue UA research in academia (University of Guelph, University of Rosario), and their work continues to influence and develop municipal and national policies on UA. In other instances, UA has been incorporated as part of the work of international organizations such as the World Food Program. Others continue to be advocates in the NGO sector and with community-based organizations.

In 2005, IDRC published *AGROPOLIS: The Social, Political and Environmental Dimensions of Urban Agriculture*, which summarized the work of nine AGROPOLIS awardees. Due both to demand plus the availability of many excellent unpublished reports, need for a second volume became evident. This book describes the research of more recent AGROPOLIS awardees on topics ranging from food security, livelihoods and community building.

AGROPOLIS has now evolved into ECOPOLIS – a new programme that will build on the former programme's success. ECOPOLIS drives IDRC into a new direction: support for design. ECOPOLIS will continue to fund scholarly research, but it will also encourage engineers, architects and planners to create and build innovative solutions to poverty in cities. IDRC's *Cities Feeding People* programme has evolved into a broader urban poverty and environment *problematique*. In addition to urban agriculture, ECOPOLIS funds research on water and sanitation, solid waste management, vulnerability and land tenure.

IDRC is proud to have supported AGROPOLIS and is convinced that funding graduate research is an exceptional investment that unleashes the knowledge, creativity and enthusiasm of young researchers. We are very proud of our AGROPOLIS awardees and hope that you will find their research as enlightening as we have.

Naser I. Faruqui
Program Leader, Urban Poverty and Environment
September 2008, Ottawa

Acknowledgements

The AGROPOLIS awards programme is the result of the efforts of many individuals and the support of many institutions. Originally conceived by the Cities Feeding People (CFP) programme of Canada's International Development Research Centre (IDRC), AGROPOLIS continues within a new programme initiative created in 2005, Urban Poverty and Environment. As the team leader of CFP, Dr Luc Mougeot was the main voice that drove IDRC in the direction of urban agriculture (UA) in the mid-1990s. He deserves great credit for his efforts and vision.

The CFP programme team comprised Dr Ola Smith, Naser Faruqui, Dr Daniel Buckles, Dr Ana Boischio, Brenda-Lee Wilson, Kristina Taboulchanas and Karen Trebert. Wendy Storey and Liliane Castets-Poupart managed the daily functioning of AGROPOLIS and became a welcome source of support and advice for many of the awardees.

AGROPOLIS awards were adjudicated by a committee of international professionals who have dedicated significant portions of their careers to furthering the field of UA. Over the years, the following individuals provided their insights and overviews to the programme and ensured its rigour: Dr Beatriz Canabel Cristiani, Dr Chris Furedy, Dr Diana Lee-Smith, Professor Godfrey Mudimu, Dr Donald Cole, Dr Tony Binns, Dr Paule Moustier, Dr David Midmore, Dr Raphael Yuen, Dr Juan Izquierdo, Dr Barry Shapiro and Dr Daniel Buckles.

The awardees have received the support of a large number of academic supervisors who steered the daily research activities of every awardee. While too numerous to name, these individuals have acted as champions of UA within their institutions. In addition, they continue to work with students to improve their research skills and to encourage an ongoing interest in UA.

Dr Donald Cole, Dr Joe Nasr, Dr Nancy Keranja and Dr Guéladio Cissé are particularly thanked for their role in helping to provide editorial advice on the content of the chapters, while Eric and Katharine Fletcher aided in the style editing. I am grateful for their help and collegiality. IDRC's Urban Poverty and Environment (UPE) team, in particular Naser Faruqui, has been supportive of our awards programmes from the outset and has been available to discuss and to offer sound advice. On an ongoing basis, Nicole Mayer assists our entire team and also helped a great deal with this book.

Finally, without the help of three colleagues in particular, this book would not have been possible. First, I must thank Alison Clegg and Anne-Marie Legault, whose organizational efforts kept many things moving along.

They ensured that the 't's' were crossed, and the 'i's' dotted. Bill Carman, who manages IDRC publications, has always a sober and well-informed perspective and, in fact, offered the initial suggestion that a second volume of AGROPOLIS papers would be a welcome addition to the literature on the topic.

Mark Redwood
Senior Program Officer, UPE
September 2008

List of Acronyms and Abbreviations

AAS	atomic absorption spectroscopy
ACSA	Asociación Comunitario San Augusto
ADI	acceptable daily intake
aeu	adult equivalent unit
AGROPOLIS	International Graduate Research Awards in Urban Agriculture
ANCAR	Agence Nationale de Conseils Agricoles et Ruraux (National Agency for Advice in Rural Extension)
APHA	American Public Health Association
AREX	Agricultural Research and Extension Services
ATE	average time to emergence
AWWA	American Water Works Association
BRIC	Brazil, Russia, India, China
CBO	community-based organization
CEPIS	Centro Panamericano de Ingeniería Sanitaria y Ciencias del Ambiente
CGIAR	Consultative Group on International Agricultural Research
CILSS	Comité Permanent Inter-états de Lutte Contre la Secheresse (Permanent Inter-state Committee on Drought)
CIP	International Potato Center
CIPSTAT	CIP Statistical Analyser
COD	chemical oxygen demand
COPACEN	Coopérative Agricole du Centre de N'djili
CPWF	Challenge Program on Water and Food
CVM	contingent valuation method
DDT	dichloro-diphenyl-trichloroethane
DIGESA	Environmental Health Directorate of the Ministry of Health
FAO	Food and Agriculture Organization
FBO	faith-based organization
FC	fecal coliforms
FISE	Fondo de Inversión Social de Emergencia
FRN	Federal Republic of Nigeria
FUNDECI	Fundación Nicaragüense Pro-Desarrollo Comunitario Integral
FW	faucet water
GIE	Groupement d'Intérêt Economique (Economic Interest Group)

GIS	geographical information system
GMB	Grain Marketing Board
GOAN	Ghana Organic Agriculture Network
GPS	global positioning system
GRA	government reservation area
HCB	hexachlorobenzene
IAGU	Institut Africain de Gestion Urbaine
ICMSF	International Commission on Microbiological Specifications for Food
IDRC	International Development Research Centre
IDW	inverse distance weighted
IFAN	Institut Fondamental de l'Afrique Noire
INTA	Instituto de Tecnología Agropecuaria
IPM	integrated pest management
IWMI	International Water Management Institute
JUR	Junta de Usuarios del Rio Rímac
KMA	Kumasi Metropolitan Assembly
KNUST	Kwame Nkrumah University of Science and Technology
LSD	least significance difference
MECD	El Ministerio de Educación, Cultura y Deportes
MEPA	multi-actor ecosystem participation approach
MoFA	Ministry of Food and Agriculture
MPN	most probable number
MRL	maximum residue limits
NALEP	National Livestock Extension Programme (Government of Kenya intervention to help urban farmers)
NGOs	non-governmental organizations
NPK	nitrogen (N), phosphorus (P) and potassium (K)
PAMA	Programa de Adecuación y Manejo Ambiental (Environmental Management Program)
PANS	participatory approach to nutrition security
PCQ	Projet de Compostière de Quartier
PEAR	participatory education and action research
PLA	participatory learning and action
RDA	recommended dietary allowances
SANDEC	Swiss Federal Institute for Water and Sanitation in Developing Countries
SAPs	structural adjustment programmes
SEDAPAL	Drinking Water and Sewerage Service of Lima
SENAHUP	National Support Service for Urban and Peri-urban Horticulture
SLU	sludge
SPSS	Statistical Package for Social Sciences
SRI	Soil Research Institute
SSHRC	Social Science and Humanities Research Council
UA	urban agriculture

UCOOPMAKIN Union of Market Garden Cooperatives of Kinshasa
UNDP   United Nations Development Programme
UPA   Urban and Peri-urban Agriculture
UPAL   Urban and Peri-urban Agriculture and Livestock
UPROVAN   Union des Producteurs de la Vallée des Niayes (Producers
            Union of the Niayes Valley)
WEF   Water Environment Federation
WHO   World Health Organization
WTP   willingness to pay
WW   wastewater

Introduction

Mark Redwood

Urban agriculture (UA) has long been dismissed as a fringe activity that has no place in cities; however, its potential is beginning to be realized. In fact, UA is about food self reliance: it involves creating work and is a reaction to food insecurity, particularly for the poor. Contrary to what many believe, UA is found in every city, where it is sometimes hidden, sometimes obvious. If one looks carefully, few spaces in a major city are unused. Valuable vacant land rarely sits idle and is often taken over – either formally, or informally – and made productive. Urban agriculture is a long-established livelihood activity that occurs at all scales, from the small family-held market garden to the large agri-business located on the fringe of a city. It supplies food to the city and income to those who farm. Above all, UA is making an important contribution to food security for those who do not have easy access. In essence, UA is the true realization of the statement that 'necessity is the mother of invention'.

In the 21st century, food comes with baggage. Mechanized farming and the increased yields associated with fertilizer and pesticide usage have reduced employment. Accordingly, farmers are relocating to cities in search of work. As wealth spreads, appetites change, and food is travelling further and further from where it is produced as people demand specialty goods. While food choices increase for the wealthy few, others are exposed to nutrition and health risks because of their lack of secure food sources. Market changes associated with biofuels, high oil prices and inflation are raising the cost of basic goods, which leads people to seek alternative ways to secure their food.

Meanwhile, the historic separation of the uses of 'urban' land from what has traditionally been considered 'rural' has relegated UA to a position of being a minor economic sector at best or irrelevant at worst. In general, policy has followed suit. Many cities, for different reasons, have ignored the contribution of UA and settled on disingenuous prohibition of the activity.

But this is changing for the better, since acceptance of UA is growing in many municipalities (Mougeot, 2006; Van Veenhuizen, 2006).

The fourth World Urban Forum in 2006 showcased the crucial importance of UA in cities of the 21st century. During the forum, statistics were presented which show that by 2006 more than 50 per cent of the world's population is living in urban areas. Moreover, projections indicate that by 2050 it is expected that two-thirds of humanity will live in urban areas. Thus, the Forum confronted delegates with the challenges of such a rapid and historic change in human geography. The Forum was also notable for bringing UA in from the fringe and introducing it during a major international event whose attendees included mayors, government ministers, international organizations, researchers and members of civil society. Urban agriculture was the main topic in a number of networking events, product launches and in the booths of at least 20 institutional partners and eight cities. Around 1000 delegates attended the networking events that took place specifically on UA. The acknowledgement of UA and its presence at such a major event is indicative of wide changes that are taking place with regard to the politics of how cities are viewed and how the value of land – and food production – is perceived.

Enter AGROPOLIS

Between 1998 and 2005, the AGROPOLIS programme of the International Development Research Centre (IDRC) supported graduate research on UA. In 2005, *AGROPOLIS: The Social, Political and Environmental Dimensions of Urban Agriculture* (Mougeot, 2005) was published. This book contained ten chapters of primary research conducted by graduate students who did their research with support from IDRC. Much of the work presented in *AGROPOLIS* was unique because, for the first time, an audience was exposed to baseline work on UA in many parts of the world that had not previously been studied. Now, with more than 60 awardees and significantly more work to reveal and examine, IDRC has decided to publish this second volume.

The appeal of UA is that it is a subject that crosses many disciplines. In an age where research is challenged to deliver results based on an integrated approach to research, UA offers an uncommon breadth. This volume, for instance, contains research from engineers, agricultural scientists, sociologists, planners and geographers. It is IDRC's view that participatory research, done in an environment that balances biophysical and social approaches, achieves aims that are beyond those of singularly focused research. This is indeed a challenge, and one that takes practice, but the effort is worthwhile. Also, in keeping with the spirit of *AGROPOLIS*, each researcher in this second volume had an association with a development group in order to increase the likelihood of the practical application of their work.

This introduction provides an overview of the chapters in this book and organizes them around key themes associated with UA. The first section represents a general background explaining the crises that the rise of cities presents – along with the opportunities. This is followed by a review of how UA represents a valid way to address and minimize the difficulties marginalized people face on a daily basis. Therefore, this introduction presents a synthesis, while the chapters that follow offer a more substantive look at each case.

THE RISE OF CITIES AND CITY FARMING

No longer isolated nodes of intense economic development and human settlements, many cities are now growing into large, continuous urban corridors containing millions of people. In cities, questions related to livelihoods and society are complex. In cities in developing countries, migration from rural areas is still the primary source of people. However, these days a second generation of children of migrants is increasing urban populations, making the competition for scarce resources even more fierce. Poverty in the fringe of cities and people being pushed into confined spaces inside cities has led to a situation where one in six people on the planet is living in a slum (UN-HABITAT, 2003), a fact that Davis (2006) uses to justify what he calls the 'Planet of slums'.

The pace of urbanization is unprecedented. Economic migrants seek better opportunities and jobs in cities while many rural economies stagnate. The trend of urbanization (see Figure 1) is most notable in the rapidly expanding economies of China and India, as well as in areas of the world where there are significant population and environmental pressures without much economic growth. The Gulf of Guinea along the coast of West Africa, for instance, represents an extreme example of massive urbanization. In 1960, there were 17 cities with a population of more than 100,000; now there are more than 300 cities of that size (Davis, 2006).

In the 1950s and 1960s, the drive to modernize economies in the south shifted the onus of economic activity onto the city. An important social result of this trend was the attraction of people to places of investment and employment, which resulted in an out-migration from rural areas. Seen by liberal development planners as the way out of poverty, the transition of emphasis to urban development and the encouragement of industrial economic activity concentrated on cities. This became known as the 'urban bias'.

In the 1970s and 1980s, a competing strategy became more popular, and development planning shifted to supporting rural livelihoods and decentralizing development efforts into smaller cities and towns. These towns were seen as important market centres for food and other goods produced in rural areas. Despite such efforts to revitalize rural areas, the pace of urbanization remains rapid. The sheer numbers of rural people moving to cities, and the growth of city populations due to high birth rates, continues to strain natural resources and the capacity of governments and states to cope.

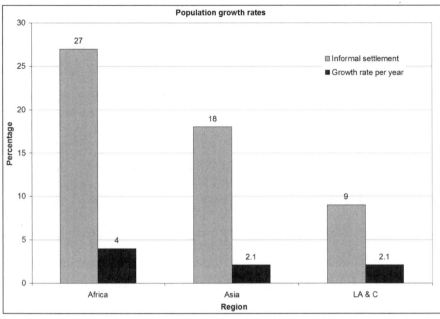

Source: WHO, 2000

Figure 1 *Urban growth rates compared with the expansion of informal settlements in African cities*

Settlements like Kieran (Nairobi), Pinkie (Dakar), Mainsheet Knars (Cairo) and Villa Maria Triumph (Lima) are testimony to the impact of rural to urban migration. As Figure 1 shows, the growth of unserviced areas, or 'slums', has far outpaced the overall growth of cities.

Many who move to urban areas do not find the jobs and opportunities they seek. Therefore, adopting UA is a common survival strategy used by the poor not only to deal with food insecurity and poverty, but also to organize with fellow citizens and improve the quality of life in their communities. Urban growth has also concentrated food demand in cities. Smit et al (1996) estimate that 15–20 per cent of the global food output is grown in cities.

In times of geopolitical or economic crisis, UA has become one way to secure food. As it turns out, the macroeconomic climate has been a significant influence in upsetting food security throughout the world. Structural adjustment programmes (SAPs) in the 1980s led to widespread currency devaluations, price increases for basic goods and the removal of subsidies for food production. In their effort to stimulate economic growth, SAPs actually removed some of the critical lifelines of the poor, thus shifting the focus of some economic planners onto the informal economy – including UA – as the poor discovered ways to survive. More recent economic crises such as happened in Argentina in 1999–2002 dramatically altered the landscape of employment and encouraged a growth in UA accompanied by policy support, as evidenced by the Pirouette programme of the federal government and city of Rosario. Another example of thriving UA is the Gaza Strip: with 3600 people per square kilometre, it is one of the world's most densely populated regions, with limited access to food imports.

Defining urban agriculture

Where there are people there is a market for food. In the eyes of planners, architects, politicians and developers, trained as many across the world are, in the arts of town planning by European colonial systems, farming in a city was considered a practice either to be discouraged or ignored. Meanwhile, UA thrived in confined spaces because people in search of a source of income who have access to land and water will practise it, regardless of restrictive policies.

Because of the history of agriculture in the European context, many people consider agriculture and the city as separate and distinct, but this has never reflected reality. Urban agriculture has always existed. Just as settlement patterns rely on good arable land in order to secure a nearby productive food source, as cities grow, their footprint paves over agricultural land. Therefore, a city's expansion de facto means there is an ongoing need for more sources of food.

The classic and widely used definition of UA comes from Mougeot (2000):

> *Urban Agriculture is an industry located within, or on the fringe of a town, a city or a metropolis, which grows and raises, processes and distributes a diversity of food and non-food products, (re)using largely human and material resources, products and services found in and around that urban area, and in turn supplying human and materials resources, products and services largely to that urban area. (p10)*

This definition links confined space production, related economic activity, location, destination markets (or home consumption) and the types of products produced in a dynamic interaction that can vary from one urban area to another. The breadth of this definition has not been challenged, and it influences the extent of research on the subject. Research, and increasingly policy, is now acknowledging that peri-urban and urban agricultural systems operate in a very different context than rural systems. Urban agriculture not only presents research conundrums associated with natural science (agronomy, pollution, water and soil quality), but also important questions of a social nature (land markets, rural to urban migration). Actually, UA research is needed so as to study policy and 'technocratic' responses in the form of planning, law and legislation.

Urban agriculture also nicely bridges the gap associated with the unfortunate tendency for development research and practice to gravitate towards two poles, 'urban' vs 'rural'. This polarization has missed some of the important nuances that occur in what some are calling the 'peri-urban interface' (Allen, 2003). This notion is important to our understanding of UA since it broadens the notion to include those distinctly non-rural agricultural systems located on the outskirts of cities. In fact, the term 'UA' has now expanded to include peri-urban areas as part of its context. The reader will note that in this volume several chapters draw on both 'UA' and 'peri-urban agriculture'. This is because some authors use the terms to differentiate forms of UA that take place in the downtown core rather than the periphery of cities.

Why urban agriculture?

Perhaps the most dramatic argument in favour of UA comes from existing data on the proportion of income that city dwellers spend on food. Table 1 illustrates the need to find more reasonable sources of food and provides a strong argument for UA as a household supplement that can counteract the worst effects of poverty. Its effectiveness is not limited to poverty reduction at the household level: it also creates economic spin-off industries and employment, plus it improves the urban biophysical environment (Moskow, 1999).

Urban agriculture also acts as a catalyst for political organization. A survey of producer groups in 2005–2007 identified that organizations based around UA play a significant role in social cohesion, offering technical training and providing a platform for political lobbying. Such groups have successfully lobbied for municipal policy change (Amsterdam), acted as the voice of farmers to lobby for recognition (Dakar, Villa Maria Triunfo) or provided technical assistance (Montreal) (Santanderau and Castro, 2007).

The administrative limits of a city also play a role in determining the extent of the practice. IDRC-supported work in Latin America demonstrated that within the city limits in that region's cities, there are large areas of vacant land. For instance, in Quito, Ecuador, 35 per cent of city land is vacant and often being used for agriculture (2001 data). In Rosario (2003 data), the amount is 80 per cent (IDRC, 2004). Recent data from Abomey

Table 1 *Percentage of income spent on food by low-income residents in selected cities*

City	Income spent on food (%)
Bangkok (Thailand)	60
La Florida (Chile)	50
Nairobi (Kenya)	40–60
Dar es Salaam (Tanzania)	85
Kinshasa (Congo)	60
Bamako (Mali)	32–64
Urban USA	9–15

Source: Akinbamijo et al (2002)

and Bohicon, two cities located in Benin, West Africa, shows that agriculture is the main activity for 3–7 per cent of people living in their downtown core. However, 6 km from the city limits, in the peri-urban area, the percentage grows to 50 (Floquet et al, 2005). In the five urban districts that make up downtown Hanoi, 17.7 per cent of land is used for agriculture (Mubarik et al, 2005).

The status quo response of governments in their reaction to UA has tended to be to prohibit the practice. Often this is a policy that stems from simply regarding UA as a form of resistance to urban development priorities as determined by planners. Some cities have, by virtue of being exposed to UA and farmer groups, changed their perspective and put in place systems that are designed to support UA, or at least remove the most draconian restrictions on the activity. However, even when rules are in place, they are often not well understood or enforced. In this volume, Mutonodzo points out that in Harare 40 per cent of the people practising UA were unfamiliar with any laws related to it. Moreover, one in five considered the existing legislation to be hostile towards the practice.

Nonetheless, progress is being made. The number of municipalities that have policies in favour of UA has increased dramatically in recent years. Accra, Beijing, Brasilia, Buluwayo (Zimbabwe), Governador Valdares (Brazil), Havana, Hyderabad, Kampala, Rosario (Argentina) and Nairobi are a short list of a growing number of cities that are being proactive on the topic. Another popular way of supporting urban farming has been food-policy councils. These represent an increasingly common way of bridging community groups with municipal politicians and bureaucrats. Amsterdam, Toronto, Vancouver, London, Detroit and Pittsburgh all have councils that encourage locally-based food systems.

The efforts of researchers, farmers and other institutions have been built upon by visionary political leaders. In Accra, for instance, UA policy has been developed and the deputy minister of food and agriculture has signed a statement of vision on UA, and the government is giving awards to UA farmers to help implement better practices associated with UA. In 2003, the Harare Declaration was signed. Represented by ministers of local government and agriculture, it commits their intention to develop UA in five African countries (Kenya, Malawi, Swaziland, Tanzania and Zimbabwe).

The Food and Agriculture Organization (FAO) now readily accepts UA has the potential to strengthen urban food security. A project of FAO conducted with IDRC worked with farmer organizations in Phnom Penh, Harare and Cairo amongst other cities to develop first-hand knowledge of how farmers organize around UA to lobby for their rights to land, water and safe food (FAO, 2007). These groups often exist as a reaction to the pressure of development and land markets that presses UA to the margins.

It has taken some time to develop appropriate policy responses, but where city governments are supportive the reaction has been very strong. The development of the 'farm to fork' approach by the International Water Management Institute (IWMI) is helping policy makers realize where interventions are most useful (IWMI, 2008). By analysing the chain of production through the lens of the farm-market-consumer continuum, policy makers can apply specific targeted interventions with maximum impact. For instance, interventions at the market level are an important policy entry point for municipalities. Consider, for example, the results from Soto in this volume that show an *increase* in microbiological risks from washing vegetables with dirty water. This helps us realize the importance of well-managed health campaigns that articulate best health management practices from the production of food through to the market and onto the consumer.

Urban agriculture, once a fringe topic, could now arguably be considered a field in and of itself due to the combined efforts of numerous organizations. What is most evident is that UA has become popular mostly because it makes enormous sense to many people. Some disagreement on the influence of UA still exists, but the logic that food will naturally be grown close to the market for that food is not debated. The contents of this book add fuel to the concept that UA, as a livelihood strategy, will continue to exist as long as the demand for food exists. Moreover, the importance of UA is such that it should not be left in a policy vacuum. Its positive and negative attributes need to be confronted by well-informed policies and programmes.

OVERVIEW OF THIS BOOK

The content of this book stems from primary research conducted by AGROPOLIS awardees in Africa and Latin America between 2002 and 2005. In Africa, research was conducted in Ghana, Senegal, Malawi and Nigeria. These chapters present work on urban food systems, urban planning, the use of participatory methodologies and wastewater use. Areas where the amount of research is relatively limited such as Zimbabwe and the Democratic Republic of Congo are included. A study in Kenya focused on household food security and nutrition in a poor area of Nairobi. In Latin America, two papers explore the use of UA and wastewater in Lima, Peru. A chapter written from work in Nicaragua (Managua) explores social ecologies and networks though a review of home gardening practices, while another chapter presents initial findings related to the health of farmers in peri-urban Rosario (Argentina).

In order to maximize the possible impact of research, the AGROPOLIS programme insisted that awardees associate their work with ongoing action, programmes and projects in the field, linking their research with practice. This type of relationship acknowledges the role UA can play in addressing poverty, while recognizing UA practices must minimize some negative environmental and health impacts. Urban agriculture is only now entering the mainstream and support for policy development is still required. By encouraging work on UA, these researchers are encouraging UA to be recognized as a viable commercial activity, thus reducing the possibility that its practitioners will remain marginalized.

Access to resources: Land and water

Not surprisingly, the environmental inputs required for UA – land and water – are the focus of a number of the papers presented here. Both resources are scarce in cities and a lack of tenure and legal status forces poor farmers to work on marginal lands with water of low quality. Indeed, wastewater has attracted a great deal of attention in recent years (Faruqui et al, 2004; Redwood, 2004; IWMI, 2008). Considered valuable by farmers for its nutrient properties, raw or poorly treated wastewater poses a significant health risk to those that use it; therefore, it has often been banned. In his chapter on wastewater use in Ghana, Amoah points out that banning its use is now accepted as a failed policy. Soto's research also illustrates that farmers rely on observed phenomena when assessing risk. In his study of farmers in the Rímac River watershed, Soto notes 75 per cent of the farmers surveyed rated domestic sewerage as a significant problem – something they see every day. Meanwhile, only 11 per cent considered mining a major concern, despite the prevalence of that industry and related pollution in the watershed. In her analysis of wastewater farmers in Dakar, Senegal, Chaudhuri observes that farmers in the area she studied were relatively aware of the dangers posed by pesticides. However, they were not concerned with pathogenic risks posed by wastewater despite the fact that many farmers exhibited illnesses associated with water-borne diseases.

Meanwhile, urban development and resulting pressure on the land is changing the nature and function of UA. Access to land is limited by increasing urban sprawl and upward pressure on land markets. In Lima, Maldonado documents a 26 per cent reduction in land available for agriculture in Carapongo, a suburb of Peru, between 2002 and 2006. During this same period, the amount of land used for housing increased by 75 per cent, placing a strain on farmers, particularly those who do not own their own property. Using statistical data from Zaria, Nsangu shows that UA was most often present on sites where planned developments had not yet occurred. In fact, he concludes that most of the land in Zaria's master plan (year 2000) is actually being used for agriculture regardless of the original intent. In this case, UA is happening on lands that are designated for development, where construction has not yet started. As happens with many cities, the ambitious plans of politicians, architects and planners remain unfulfilled, allowing UA to fill the void.

The intensity of UA as a result of growing in limited space places stress on soil fertility and leads to the reuse of liquid and solid wastes as a way to capture nutrient value (Mougeot, 2006). These wastes possess an economic value which is not captured by formal markets and, therefore, offer a potential source of revenue to those who can provide them. This 'closing' of the nutrient cycle also serves to reduce the amount of waste destined for landfills or along roadsides. Several writers (Amoah; Eriksen-Hamel and Danso) tackle this issue of nutrient recycling and waste use.

In Kumasi, Ghana, research documented farmers' willingness to pay (WTP) for both compost and poultry waste as fertilizer (Eriksen-Hamel and

Danso). These authors found that both compost and poultry manure are highly valued by farmers. The benefit of the use of compost is supplied by increased yields: for instance, both tomato and lettuce yields increase by 12 per cent when compost is applied. In terms of poultry manure, its inherently high nutrient content makes the potential development of a natural fertilizer market very appealing. They also note that because chemical fertilizers are much more expensive, farmers were willing to pay a considerable amount for poultry waste. Also, according to the WTP survey, compost fetches a price of up to US$3/50 kg amongst vegetable farmers. This proves that, in the right quantities, compost can be supplied for profit. In urban areas, farmers are more reliant on highly valued and plentiful poultry manure – as an example, there are 300 poultry farms in Kumasi.

The same chapter on Kumasi, Ghana, emphasizes that the importance of research on these economic questions cannot be understated. If economic benefits are made clear, farmers will opt for what they see as the most beneficial path. In fact, they will employ production methods even if there may be associated health risks. Therefore, it is important for a policy approach to have the twin objectives of enhancing economic benefits along the chain of UA production, while also being creative in regards to how to promote hygiene considering that health risks are often imperceptible to farmers.

Gender analysis, participation and innovation in methodological approaches

Development research has also evolved to a point where subjective work can be ordered in a systematic way to enhance more empirical, positivist methods (Thomas and Mohan, 2007). Some traditional methods of qualitative and quantitative enquiry have worked well in UA research, but increasingly, analysis associated with power, ethical and moral issues is being used to corroborate positivist analysis. Being such a broad topic, UA requires variety in the types of methods used in enquiry. AGROPOLIS awardees were encouraged to make 'action research' the cornerstone of their methodology and approach. The premise of this was that universities and research institutes working in partnership with non-governmental organizations (NGOs) and community-based organizations (CBOs) tend to support the link between the generation of new knowledge and the field implementation of that new knowledge. In fact, however, neither of the extremes are ideal as stand alone approaches. Research can benefit a great deal when scientists ask questions about political ecology, while social analysts attempts to include more scientific rigour.

In the first book presenting AGROPOLIS awardees' work, Mougeot (2005) noted that awardees were working at different places in the research cycle. Some researchers were presenting baseline work on UA because their cities had only performed limited work, previously. In this book, several chapters are in locations where information on UA is basic at best, or non-existent – Zaria and Kinshasa are examples of this. Meanwhile, other chapters are building on an already considerable knowledge base, as is the case in Kumasi, Rosario and Harare where baseline research has already been done. As a result, the methodology in each chapter differs and we are exposed to a broad range of analysis. Standard quantitative methods are used in some of the work while in others, participatory methods are employed to explore a very specific aspect of UA. On one hand, having a systematic approach across all of the research would have yielded some strong comparative evidence; however, the richness of this breadth of approaches illuminates different learning within each context.

Ten years ago, hardly any data existed on UA, such that traditional forms of enquiry were employed to generate baseline data. Now, the field has moved into a new phase where more sophisticated methods are employed and research is able to answer the questions raised by earlier work. In this book, traditional methods of enquiry are interspersed with some attempts at new methodological approaches. This section of the introduction attempts to tease out some lessons from this variety.

Participatory learning and action

One clear trend in development research in the past two decades has been the rise of popularity of participatory methods in social science research. At its most basic, participatory research implies that researchers are engaging with 'beneficiaries' in a way that will lead to an empowerment of that group. Participatory research also implies that the group researched will have more ownership of the project and its results and is notable as a way that research can empower communities and not simply be a process of extraction.

Nevertheless, there are voices that critique the approach as overly reliant on information-gathering from sources that are already powerful and entrenched politically (Thomas and Mohan, 2007). This critique also extends the fact that in the participatory research process, the researcher has significant leeway to subjectively interpret results. However, this can be very misleading as local knowledge is not infallible or without its own political baggage. For this reason two things are necessary in participatory research. First, ensuring that the participatory research process is done in a *systematic* way, and second, to ensure that results using participatory techniques are triangulated with results acquired using other methods. To work systematically, researchers should be active in developing their background research, typologies, categories, interview questions and so on in advance, but should incorporate some flexibility in the process to adapt as required.

Critiques aside, one advantage of PLA is that it provides a context for the use of other analysis such as quantitative methods that are used in biophysical research. Also, PLA can help to ensure that the right questions are being asked. In terms of UA, the following are examples of particularly relevant questions: Who is growing what and why? Are the concerns of the local people the same as those of the researcher? What economic factors influence their decisions? Which institutions are involved and what role do they play?

Participatory education and action research

Most of the research conducted in the AGROPOLIS programme included in this book uses PLA or a variant of this method such as participatory education and action research (PEAR). Chaudhuri's chapter is devoted to the presentation of PEAR as a method of uncovering health-related information. PEAR requires a high degree of involvement on the part of the researcher, in order to untangle power dynamics and lay the foundation for advocacy on the part of the farmers' group with whom she worked. While this experiment can yield very interesting results, assessing the transformative nature of the researchers' involvement is more difficult.

Others made use of gender-disaggregated participatory workshops to explore the role of women in UA (Soto), to develop a historical timeline of the community, for mapping (Shillington) and to analyse feasible alternatives to UA for income diversification.

Notwithstanding the critique, small investments in participatory research can make a difference. Njogu's work is an example. In her chapter, she illustrates that her research involved not only a baseline but also an intervention in a subset of households within her sample. Her activities – an attempt at raising dietary diversity – actually led to an increase in the number of crops consumed with whom she worked. For calories, protein, vitamin A and iron, there was a statistically significant increase in the number of children achieving the recommended daily allowance.

The livelihoods framework

Linking research with policy requires that research be multidisciplinary and involve as many stakeholders as possible. As a way of tackling the multiple stakeholders and disciplines, several authors used the livelihoods framework in developing their research protocol. The basis of the framework is analysis that is people-centred and looks at structural issues such as access to resources, human capacity, social and political assets (Buelcher, 2004). It is designed to capture the interplay between development, environment and livelihoods.

In this volume, some authors use the approach (Soto, Maldonado) as a way to capture information using a diverse set of methods and stakeholders. In Maldonado's work, the livelihoods approach was important in teasing out detailed information about the motivation and situation of farmers in Carapongo, Lima. Maldonado is right to have identified self-perception and esteem as important elements of human capital that can impede (or facilitate) development. As part of his research, he asked farmers about their self perception. Perhaps not surprisingly, a lack of land was cited by one farmer as the main reason for having a 'poor' perception of oneself. On the other hand, access to basic services and other sources of income were reasons for an improved self perception. It will come as no surprise to those who have worked in field research that few participants described themselves as 'poor'. Although the sample was small, the effort to explore self-esteem amongst farmers suggests that future research can consider this as an enabling factor when designing strategies in a participatory manner of how to deal with problems associated with UA.

One lesson that can be drawn from the work involving the livelihoods approach in this book is the effective support it can offer a researcher trying to categorize multiple sources of information. By disaggregating our understanding of poverty into the five different categories of assets, it becomes a flexible and detailed framework where both quantitative and qualitative information can fit. Using a livelihoods framework also allows a researcher to corroborate learning and needs analysis associated with different stakeholders.

Gender

It is well known that women and children are important actors in UA in terms of production and marketing as well as in associated composting and recycling activities. Women also face problems in accessing some services such as credit and extension and training services, and also in access to legal or customary land tenure – in fact, anything associated with asset ownership (Hovorka and Lee-Smith, 2006). In research, gender-disaggregated data is frequently provided, but often with little actual analysis.

In this volume, several authors present data on gender and UA. Mkwambisi did some focused analysis with female-headed households. He suggests these homes are more efficient farms and notes that women are highly represented as UA farmers. In the study conducted on UA in Harare, the dependency ratio (defined as the proportion of adults to dependants, young and old) was found to be significantly higher in female-headed households than in male-headed households, corroborating earlier evidence from sub-Saharan Africa SSA (Maxwell et al, 2000). This suggests more challenges in achieving food security in households headed by women as they are expected to cope with more mouths to feed.

Evidence presented in this book also highlights improved efficiency in households practising UA headed by women. Female-headed households are more efficient at farming (Mkwambisi) and also over-represented as UA farmers. Maldonado's work in Lima found that on average, women spend seven hours working in agricultural activities and another eight hours in household activities, whereas men spent nine hours working only in agricultural activities. In addition, Maldonado found that women are more present in the marketing of UA produce. Thus, the labour-intensive activity of urban farming is closely associated with the classic model of the 'double day'. Mkwambisi demonstrates that female-headed households and poorer households benefit significantly more in terms of financial gain than wealthier households for whom UA is considered more of a 'hobby'.

Gender analysis requires asking questions that relate to social and cultural norms and behaviours that are not always adequately captured by scientific enquiry. Shillington's work presents the notion of 'social ecologies' through which men and women interact. In her mapping of these ecologies in patio and balcony gardens in Managua, she found that women identified and used a greater diversity of species than men, suggesting that they pay more attention to the patios. As she puts it, 'patios tended to reflect women's desires and needs'. This corroborates gender analysis of women being more active in private spaces in contrast to public ones.

Constraints facing researchers

In some parts of the South, conducting research is becoming increasingly difficult. More and more people are unwilling to participate since research too often is not seen to lead to a tangible change in one's livelihood. The distance between scientists and researchers is often huge and difficult to surmount (Bentley, 1994). Stories of researchers making promises that are never fulfilled abound. There is a common expectation that study will lead to change, and while this is sometimes the case, it is not always true. Njogu describes challenges encountered in her follow-up survey where some of those who had participated in the baseline analysis refused to be interviewed in the follow-up, because they had not been selected for the pilot intervention.

The challenge of gathering data on incomes and financial information is well known and was mentioned by a number of the authors. It was also suspected in some cases that some respondents reported lower incomes than they actually received in hopes that the researcher might be able to facilitate access to government programmes and subsidies. In his chapter on Lima, Maldonado points out that during his baseline survey of income sources related to their sources of employment, accessing people's true incomes was difficult as many are hesitant for reasons of privacy. This suggests two things. First, that the ethics of research need to be tightened when it comes to accessing information on peoples' financial situation and other sensitive matters. Second, it reflects the need to use better collection methods – or at least corroborate – income data, such as reviewing tax rolls, or using income-expense journals that are monitored on an ongoing basis.

A similar problem was recognized by Eriksen-Hamel and Danso in their WTP survey. To increase the accuracy of their results, these authors avoided direct questions about WTP and instead modified their approach by introducing a bidding game. In a workshop asking participants their WTP for compost in different hypothetical scenarios, the game was introduced as a way of making their conceptual notions more concrete, accurate and as close to reality as possible.

Research related to health presents those working with few resources and over a short period of time with specific challenges. Accessing health information and identifying relationships between UA activities and health can be difficult. The chapter by Propersi about peri-urban agriculture and health alludes to problems associated with gathering health information. Soto found that in order to understand the different ways that health risks might be present, the researcher needs to assess multiple data sources and perform pathway analyses to understand how contaminants reach different foods. In Propersi's case, some respondents were hesitant to discuss health problems associated with working on farms for privacy reasons plus for fear of their employment situation. In order to overcome this, Propersi corroborated her observations with visits and data from health clinics in the area of study. She also used photographic evidence to highlight certain types of practices. Cross-referencing survey data with photographic evidence can be a useful way to triangulate information.

The social and economic implications of urban agriculture

The purpose of UA is often to raise income through sale of produce or to reduce expenditure associated with the purchase of food (see Table 1). Surprisingly, despite this perhaps obvious conclusion, there have been few direct and strong economic analyses conducted related to the value of UA (Mougeot, 2005). In fact, none of the chapters in this volume actually present primary data on the financial value of UA which suggests that it is difficult information to gather. Another interpretation is that from the outset, the researchers assumed enough was already known about the income potential of UA – where assumptions such as this are common in work of this kind. In fact, the lack of strong economic data on UA is a major weakness that is impeding its entry into acceptance, particularly by economic planners.

The contribution of UA to food security varies. In the two chapters presenting quantitative evidence on this, both found that households employing UA do see an increase in food security. For instance, according to Mkwambisi's work in Malawi, 17 per cent of household food entitlements come from UA. Mutonodzo's research found that in households that practise UA, 76 per cent of daily food requirements were met. This compared to households not practising it, where 65 per cent of daily food requirements were met according to FAO standards. This suggests that urban farmers are increasing their dietary diversity and nutritional intake through UA. The implication for poverty reduction is important in several respects. For one, if restriction on UA were to be upheld, one can safely assume that the population of urban poor using UA would be worse off. Second, the burden and cost of replacing UA would fall to the state/city taxing resources which are, in many cases, already stretched. These findings on the contribution of UA to food security illustrate the value it makes to poverty reduction.

Another finding that was expressed in several papers is that the poor are more reliant on UA than the rich. Kifuani's work on the UA sector in Kinshasa illustrates the poverty of some UA farmers. In his study, he found that most (54 per cent) of the farmers' daily expenditures were less than US$0.50 per day. Mkwambisi's study in Malawi noted that while UA contributed 9 per cent of high-income households' livelihoods, it was the main livelihood activity of 42.5 per cent of low-income households. While UA tends to be concentrated among the poor in aggregate numbers, he points out that the larger economic contribution comes from wealthier farmers who operate farming businesses that are capital- and labour-intensive. This implies that policy makers have a choice to make in their policy support depending on their strategic objectives: If it is to reduce poverty, policy needs to be oriented towards poor farmers. However, if the objective is to increase aggregate food availability, it should concentrate on the businesses of wealthier farm operators.

Nutrition and environmental health: Benefits and burdens

Much has been exposed by recent work on environment and health matters and UA in recent years. Its importance is clearly illustrated by the relevance some international organizations are placing on it. For instance, two of Consultative Group on International Agricultural Research (CGIAR) system-wide initiatives (International Water Management Institute and Urban Harvest) focus on supporting research on environmental and health matters related to the reuse of waste. Moreover, WHO's new guidelines for irrigated agriculture (2006) make explicit reference to the breadth of the problem in urban areas because cities generate the most waste. Such guidelines also provide crucial epidemiological evidence, as well as management strategies, on how to safely reuse various wastes in agriculture.

Several chapters are devoted to exploring the link between UA, environment and health. Soto notes that using water from the Rímac River in Lima for irrigation represents a major source of risk for farmers. The primary concern is one that most cities are faced with: much waste is disposed directly into rivers. Fully 97 per cent of samples taken from the river were above acceptable limits for the presence of fecal coliforms (FC). Such microbiological risks can be extrapolated to other rivers that are not only repositories for waste, but also sources of water for agricultural use. Heavy metals, especially cadmium and arsenic, also constitute a health problem in the Rímac River, where their presence as a result of mining was found in vegetables grown in the watershed.

Amoah shares his research from a large body that has been developed by IWMI-Ghana on the topic of wastewater use. The level of pathogenic and pesticide contamination of produce at the farm level due to inappropriate irrigation methods is high. In produce sampled in three Ghanaian cities, both pesticides and FC levels were above regulatory limits, indicating significant health concerns. Part of the problem is that irrespective of its source such as from shallow wells, rivers or from sewerage, irrigation water is usually well above allowable limits. The author stresses that health interventions must be adapted along the entire 'farm to fork' continuum. Hygienic practices in a market, for instance, play a significant role in the quality of produce purchased and consumed 'downstream'. Importantly, however, Amoah shows that most contamination takes place at the farm and overshadows any hygienic danger at the sale point. The implicit suggestion is that effective interventions related to health are needed at different points along the food chain: at the farm level, transportation, market and home. The WHO guidelines, released in 2006, are sure to play a significant role in the policy development related to UA.

The policy response to these types of environmental problems can be a mix of basic hygiene procedures – probably the most economically efficient – combined with more comprehensive regulatory policies. In fact, in the case of the Rímac River, Soto found that historical levels of arsenic have been reduced following regulation of the use of the heavy metal by miners. But, as many already noted, the challenge for hygiene campaigns to surmount is that of a lack of any perception of risk. For example, both Soto and Chaudhuri point out that farmers are not very aware of risks that are not observable. This poses a challenge to health educators and policy makers, who must design educational programmes that educate farmers about the risks posed by using wastes in agriculture.

The 2006 WHO guidelines for wastewater use in irrigation are oriented around a multi-barrier approach that is designed to reduce risk. For example, these guidelines envision that at different stages of the production-consumption chain, barriers preventing contamination can be adopted. Such barriers can encompass treatment, irrigation methods, on-farm prevention of direct contact with polluted water and interventions at the market.

Conflicts and other constraints

While UA offers enormous potential, it is a mistake to believe that producing food in cities is free of problems. Although a number of major constraints exist, three are particularly important: access to land, access to water and the absence of comparative economic values generated by UA in the face of other activities such as land development.

Access to land

In cities, land is at a premium. The high value of urban land often makes UA at any scale particularly vulnerable, especially in terms of how the poor can access space to grow food. In the Mkwambisi chapter, most farmers in Malawi indicate that for them, access to land is the most pressing constraint on UA. Land-tenure security determines the extent of investment and self-help put into a property. Without secure tenure, farmers face eviction, uncertainty and are less willing to invest in long-term solutions related to their properties.

Access to water

Access to sources of water for irrigation causes similar problems. In the absence of safe water for irrigation, and due to the high nutrient contents present, wastewater is highly valued by farmers. As Chaudhuri discusses in her chapter, accessing raw wastewater through breaking mains and other means, with little consideration for health consequences, is a common practice in Dakar. Farmers also value this resource and have organized resistance to efforts by others to impose and enforce rules around access. In Dakar this led to the formation of the Union des Producteurs de la Vallée des Niayes (Producers' Union of the Niayes Valley, UPROVAN), a farmers' organization which actively disputes any attempt to infringe on their land by developers and others.

Absence of comparative economic values

The third major constraint is the failure to adequately measure and account for the relative contributions that UA make to alleviate poverty and contribute to a city's economy. Setting aside social and cultural 'returns' such as those outlined by Shillington in her chapter – returns that as yet have not been well measured by economists – UA does not produce the same economic value on land that would otherwise be realized as a result of housing or commercial development. The lack of high economic output and, importantly, the lack of any assets that can be used for collateral restricts access to credit and financing mechanisms that urban farmers need so as to purchase inputs and build their enterprises.

In Maldonado's study, 53 per cent of UA practitioners in Carapongo (Lima) are working on land to which they do not have title or are otherwise farming illegally. Moreover, the author found that although 14 per cent had received formal credit, few farmers use the formal banking system because they have no collateral to offer. As is the case in many unregulated or slum areas, land ownership functions within an informal system and thus credit is also acquired informally, for instance through banks, family, credit unions or NGOs. Family sources of credit seems to be the most common, probably due to the tenuous nature of UA and the perceived risks associated with offering loans to UA producers.

Other factors such as the predominance of the poor in UA, the illegal use of lands for UA, balancing health issues with social equity and the large number of women involved also contribute to the contentiousness of UA. In Kinshasa, Kifuani highlights other conflicts between various parts of the production chain in UA. Problems between producers are mostly associated with plot boundaries, access to water resources, the perception of a lack of fair competition and problems linked to agricultural equipment or inputs. Kifuani also found that 18 per cent of market gardeners have conflicts with consumers usually over prices, non-payment and various other misunderstandings. Perceptions amongst consumers can also be negative to the production of UA. For instance, those consumers who are aware of produce grown with wastewater are less inclined to purchase them or will only purchase them at lower rates.

CONCLUSION

Despite the significant challenge of being perceived as a problem, UA is now part of the debate on how to improve our cities. Policy recognition of UA is on the rise, and as it increases, so too does the level of effort amongst scientists and researchers to document the practice. Once ignored by researchers, UA now is a recognized field of research and an area that has seen some rapid policy development in the past 15 years. This is because UA supports livelihoods and generates an economic value from land that would otherwise be idle or vacant. Its economic impact reaches far beyond farmers: it stimulates employment further along the chain of production for those who market, transport and sell produce, as well as for those who provide the tools and other means of production (Mougeot, 2000). In addition, UA is a catalyst for community organization and for the greening of cities. It also increases dietary diversity because it enables a local source of fresh produce; and it plays a role in the use of solid and liquid waste.

In the first volume of work by AGROPOLIS awardees, Luc Mougeot refers to UA as the 'the oxymoron par excellence' for the simple reason that strong perceptions of what constitutes 'urban' do not include 'agriculture'. This, however, is less and less the case. The increasing acceptance of UA is a significant fact which illustrates changing attitudes towards the city as a whole. AGROPOLIS awardees have been the vanguard of research onto this topic. Many have gone on to complete their PhDs on UA, to teach and research the subject in universities, to practise it themselves, and to deliver new programmes and projects that support the activity. Although UA may always have its detractors, it is now absolutely clear that UA elicits sympathy and support. As research improves, so too will our understanding of the pitfalls and opportunities that UA can offer.

REFERENCES

Akinbamijo, O., Fall, S. and Smith, O. (eds) (2002) *Advances in Crop-Livestock Integration in West Africa*, IDRC Ottawa, Canada

Allen, A. (2003) 'Environmental planning and management of the peri-urban interface: Perspectives on an emerging field', *Environment and Urbanization*, vol 15, no 1, p135

Bentley, J. W. (1994) 'Facts, fantasies, and failures of farmer participatory research', *Agriculture and Human Values*, vol 11, no 2, pp140–150

Buechler, S. J. (2004) 'A sustainable livelihoods approach for action research on wastewater use in agriculture', in Scott, C., Faruqui, N. and Raschid-Sally, L. (eds) *Wastewater Use in Irrigated Agriculture: Confronting the Livelihood and Environmental Realities*, CABI Publishing, London

Davis, M. (2006) *Planet of Slums*, Verso Books, London

FAO (2007) 'Urban farming against hunger', www.fao.org/newsroom/en/news/2007/1000484/index.html (accessed 28 February 2008)

Faruqui, N., Niang, S. and Redwood, M. (2004) 'Untreated wastewater use in market gardens: A case study of Dakar, Senegal', in Scott, C., Faruqui, N. and Raschid-Sally, L. (eds) *Wastewater Use in Irrigated Agriculture: Confronting the Livelihood and Environmental Realities*, CABI Publishing, London

Floquet, A., Mongo, R. and Nansi, J. (2005) 'Multiple functions of agriculture in Bohicon and Abomey, Benin', *Urban Agriculture Magazine*, vol 15, pp9–10

Hovorka, A. J. and Lee-Smith, D. (2006) *Gendering the Urban Agriculture Agenda*, in Van Veenhuizen, R. (ed) *Cities Farming for the Future: Urban Agriculture for Green and Productive Cities*, International Institute for Rural Reconstruction and IDRC, Manila, Philippines

International Development Research Center (IDRC) (2004) *Optimizing the Use of Vacant Land for Urban Agriculture Through Participatory Planning Processes*, Final Technical Report, Project number 100983, IDRC, Ottawa

International Water Management Institute (IWMI) (2008) 'Agriculture, water and cities', www.lk.iwmi.org/rthemes/AgricultureWaterCities/index.asp (accessed 28 February 2008)

Maxwell, D. G., Levin, C., Armar-Klemesu, M., Ruel, M., Morris, S. and Ahiadeke, C. (2000) *Urban Livelihoods and Food and Nutrition Security in Greater Accra, Ghana*, IFPRI, Washington, DC

Moskow, A. (1999) 'Havana's self-provision gardens', *Environment and Urbanization*, vol 11, no 2, p127

Mougeot, L. J. A. (2000) 'Urban agriculture: definition, presence, potentials and risks', in Bakker, N., Dubbeling, M., Gundel, S., Sabel-Koschella, U. and de Zeeuw, H. (eds) *Growing Cities, Growing Food: Urban Agriculture on the Policy Agenda*, German Foundation for International Development (DSE), Feldafing, Germany, pp1–42

Mougeot, L. J. A. (2005) *AGROPOLIS: The Social, Political and Environmental Dimensions of Urban Agriculture*, Earthscan, London

Mougeot, L. J. A. (2006) *Growing Better Cities: Urban Agriculture for Sustainable Development*, IDRC, Ottawa

Mubarik, A., De Bon, H. and Moustier, P. (2005) 'Promoting the multifunctionality of urban and periurban agriculture in Hanoi', *Urban Agriculture Magazine*, vol 15

Redwood, M. (2004) *Wastewater Use in Urban Agriculture: Assessing Current Research and Options for National and Local Governments*, Cities Feeding People Report no 37, IDRC, Ottawa

Santanderau, A. and Castro, C. (2007) 'Social organisations of agricultural producers in Latin America and Europe: Lessons learned and challenges', *Urban Agriculture Magazine*, vol 17, pp5–7

Smit, J., Ratta, A. and Nasr, J. (1996) *Urban Agriculture: Food, Jobs and Sustainable Cities*, United Nations Development Programme (UNDP)

Thomas, A. and Mohan, G. (2007) *Research Skills for Policy and Development: How to Find Out Fast*, Sage, London

UN-HABITAT (2003) *The Challenge of Slums: Global Report on Human Settlements*, Earthscan Publications, London, UK

Van Veenhuizen, R. (ed) (2006) *Cities Farming for the Future: Urban Agriculture for Green and Productive Cities*, International Institute for Rural Reconstruction and IDRC, Manila, Philippines

World Health Organization (WHO) (2000) *Water Supply and Sanitation Assessment, Part II*, World Health Organization, http://www.who.int/water_sanitation_health/Globassessment/Global1.htm (accessed 2 January 2008)

World Health Organization (WHO) (2006) *Guidelines for the Safe Use of Wastewater, Excreta and Greywater* (vols 1–4), WHO, Geneva, Switzerland

1

Household Food Security Among Urban Farmers in Nairobi, Kenya

Eunice Wambui Njogu[1]

INTRODUCTION

Since the 1960s, residents of Kamae had lived in slum conditions, but in the year 2001 they were allocated small landholdings by the local administration of the Government of Kenya. Prior to the start of this research project, observation indicated that these landholdings could potentially increase the diversification and intensification of food-production systems. A baseline survey conducted between June and July 2005 revealed none of the 300 respondents surveyed had received any form of organized agricultural training. Therefore, no organized farming was observed. Because the extension service of the Ministry of Agriculture and Livestock and Fisheries Development was farmer-driven, only more prosperous farmers sought crop and livestock extension services. As a result, the majority of poor farmers in Nairobi manage on their own, getting no assistance or advice, which is necessary for the enhancement of their food production skills particularly in the diversification of agricultural systems (Foeken and Mwangi, 2000).

The government has been restructuring and partially privatizing the public extension system to improve extension-service delivery to all farmers regardless of their socio-economic class (Government of Kenya, 2004). However, according to Ishani and Lamba (unpublished work, 2007), urban and peri-urban agriculture and livestock (UPAL) is not recognized in this policy. Fortunately, the National Livestock Extension Programme (NALEP) has been introduced in Nairobi province to support farmers (Z. Ishani and D. Lamba, unpublished work, 2007).

Without the inclusion of UPAL as a policy issue in agriculture and livestock policies, the social and economic benefits UPAL has on vulnerable groups cannot be fully realized (Z. Ishani and D. Lamba, unpublished work, 2007). When well developed, UPAL plays an important role in cities because it improves the nutrition status of household members, generates income, provides employment and conserves the environment.

'Stunted', 'underweight' and 'wasted' are indices used by the World Health Organization (WHO) based on the measurement of height, skin thickness (fat) and age to determine the nutritional status of the subject (WHO, 1995). In most cases, children are used as they represent an indicator of the general well-being of a community based on their intake of food. Children who are wasted are considered chronically malnourished. The results of the baseline survey showed that 62 per cent of children below five years old were stunted, 53.7 per cent were underweight and 31 per cent were wasted, whether mildly, moderately or severely. Their caloric, vitamin A and iron intakes were below recommended dietary allowances (RDA) established by the WHO. The assessment of people's nutrition knowledge on such important macronutrients (carbohydrates and proteins) and micronutrients (vitamin A and iron) showed that 49.7 per cent had no idea about them. Research in Korogocho, Nairobi showed higher energy intake among urban farmers than non-farmers (Foeken and Mwangi, 2000). The exposure of this lack of agricultural and nutritional knowledge combined with the observed poor nutritional status necessitated the design and implementation of an intervention. The intervention was to introduce the concept of growing a diversity of crops and rearing small livestock to enhance households' food security.

As a result of this diversification of crops and livestock-rearing, household income increased in several ways. First, if they were marketed, the surplus food crops and livestock as well as their products fetched extra income. Second, the increased availability of food crops relieved the households' income from food purchases. Then, due to this increased income, households' food purchases became more diverse so people could incorporate higher quality foods in their diet (Bonnard, 2001). Therefore, households' food security was enhanced by increased availability and access to diversified diets.

Selected respondents were trained on how to utilize their small land-holdings to produce a diversity of crops, rear small livestock and also to teach nutrition. This was done in collaboration with the Ministry of Agriculture and Ministry of Livestock and Fisheries Development. The households were provided with the fencing material, seeds, seedlings and small livestock of their choice. All this was expected to lead to enhancement of household food security as the initial step towards poverty eradication. This paper covers the objectives, methodology and findings of the research project, in addition to offering a conclusion and recommendations based on the findings.

Objectives

The five specific objectives of the project were to:

1   Determine knowledge and attitudes of the farmers towards peri-urban agriculture (UPA) in the households.
2   Assess farming practices among the farmers in the households.

3 Determine each household's food security status in this area.
4 Assess the nutrition status of children below five years of age in the households.
5 Establish the relationship between farming practices and the household's food security.

Hypotheses

There were two hypotheses:

1 There was a positive relationship between household food security and the farming practices adopted by the peri-urban farmers of Kamae.
2 There would be a significant improvement in household food security after the intervention.

METHODOLOGY

Research design

This was an intervention study project which was semi-longitudinal in nature with an intervention period of one year. The research adopted the participatory approach to nutrition security (PANS) using the Triple-A cycle, as shown in Figure 1.1.

PANS facilitated community members to map out their problems, design their solutions and take action to implement these solutions. A baseline survey was carried out to assess the food production and food security situation in Kamae settlement. Kamae area is located north of Nairobi in Kahawa Ward,

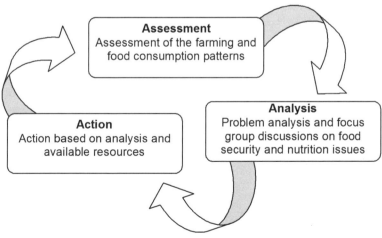

Source: Adapted from Iannotti and Gillespie, 2002

Figure 1.1 *The (PANS) Triple-A approach*

Kasarani Division. It borders the Kiambu district, which is in the central province of Kenya. Kamae, like Nairobi City, has a bimodal rainfall pattern. The long rainy season is from April to August, while the short rainy season is from late October to early December. Kamae has a total of 2036 households and a population of 9847 persons according to the 1999 Population Census (Government of Kenya, 1999).

Data were analysed to identify the food production and food and nutrition security problems. Those that were given first priority during focus-group discussions were lack of agricultural and nutrition knowledge, as well as inadequate supplies of clean, safe water. The action plan included offering education in agriculture and nutrition by the officers from the Ministry of Agriculture and Livestock and the Ministry of Fisheries Development, as well as from the researcher. Project participants each contributed approximately US$58 for the provision of water. The researcher, with support from the International Development Research Centre (IDRC), provided fencing materials, seeds and seedlings, as well as small livestock. An evaluation was done through a survey to ascertain the impact of the project.

Sample size

Three hundred households were studied during the baseline survey for situation assessment.

To obtain a representative sample, Fisher's formula was employed (Fisher et al, 1998). That is:

$$nf = \frac{n}{1 + (n \div N)} \tag{1.1}$$

where nf is the desired sample when the population is less than 10,000; n is the sample^2 when the total population is more than 10,000; and N is the estimated population of the households in Kamae area (2036). Therefore, the sample consisted of 323 households:

$$nf = \frac{384}{1 + (384 \div 2036)} = 323 \tag{1.2}$$

However, for the purpose of this study, this figure was rounded down to the nearest hundred, that is, to 300 households. Cluster sampling was employed during the baseline survey to obtain these. Every respondent was interviewed in those households within the selected cluster that contained a child below five years of age. For the sample related to the household interventions, 10 per cent of the 300 households (30) were selected randomly.

Three focus group discussions took place during the baseline survey, consisting of ten participants each, where one group consisted of men, one of women and one of youth. There was one focus group discussion during the evaluation consisting of those who were in the intervention project. Observations were made for the 300 households during baseline survey, whereas the evaluation survey included the 30 homes where interventions took place and a further 150 households. Evaluation covered 180 households because approximately 120 respondents had either moved or, for others, children were older than five years of age. Still others refused to be interviewed because they were not selected for the interventions.

Interview sessions, observations and focus group discussions
The interviews were carried out in a face-to-face situation using Kiswahili language. They were conducted during a baseline survey between June and July 2005 and at the end of the project (March–April 2006) during the evaluation process. Monitoring was continuously carried out throughout the duration of the project and observations were recorded on the observation checklist. The characteristics observed were: farming systems, labour providers, crops produced, tools used for food production, general sanitation of the household compounds and cooking facilities. Focus group discussions were held during problem analysis in September 2005 and also during evaluation on 16 April 2006. The researcher and research assistant guided and tape-recorded these sessions.

Awareness campaigns, demonstrations and intervention
To sensitize the community members on the objectives and importance of this project, an awareness campaign was held on 25 July 2005 in collaboration with the staff of the Ministry of Agriculture and Ministry of Livestock and Fisheries Development. The 30 households chosen for intervention were those that had children with high levels of wasting, and these households were divided into two groups, which were trained in two separate sessions. The first session was on organic farming while the second was on nutrition education with a bias on macronutrients (carbohydrates and proteins) and micronutrients (vitamin A and iron).

The first group was trained by the researcher and an organic farming expert in collaboration with the staff of the Ministry of Agriculture and Ministry of Livestock and Fisheries Development on 29 October 2005, followed by a demonstration on 2 November. The second group, which was trained on 16 November, was facilitated by one of the participants who had been trained in the first session, who was assisted by the researcher. The second demonstration was done on the same plot on the same day. Small livestock were kept on the two demonstration plots. The other 28 participants in the intervention were guided on how to plant crops and rear small livestock on their plots.

This study focused on the actual intake of calories, proteins, iron and vitamin A. The intakes were assessed using a 24-hour dietary recall of children who were less than five years old. Adequacy was determined by assessing those children whose intakes were below and above the RDA.

Data analysis
The quantitative data were analysed using the Statistical Package for Social Sciences (SPSS) programme. The food production systems were determined by

assessing the farming practices adopted by the households. Anthropometric data were analysed using the Nutrition Package (EpiInfo, 2000). Those children with -2 to < -1 standard deviations were classified as mildly malnourished, those with -3 to < -2 standard deviations were designated moderately malnourished, while those below -3 standard deviations were considered severely malnourished. Anthropometric measurements were compared with the US National Centre for Health Statistic/World Health Organization (NCHS/WHO) international reference standards.

Household food consumption was determined by assessing the intake of calories, protein, vitamin A and iron. The nutrient intake data were analysed using computer software (Nutrisurvey, 2004). Dietary diversity was assessed by totalling the number of food groups consumed in the households for the past 24 hours. These groups are starch (cereals and non-cereals), fresh and dry legumes and nuts, vegetables, fruits, eggs, meat, fish and milk. Households were classified as follows: those consuming 1–3 groups were considered low diversity, those consuming 4–6 groups were medium, while those consuming 7–9 were considered the high-diversity group. The level of agricultural and nutrition knowledge and the attitudes of the farmers was assessed by totalling the scores in each household.

Transcription was done using tape-recorders to determine the problems encountered in crop and small livestock production and to capture the farmers' attitudes towards urban agriculture (UA). The Pearson product moment (r) established relationships between farming practices and household food security; it tested the hypotheses at 0.05 significant levels. The t-test was used to evaluate the intervention project by comparing intakes of calories, protein, vitamin A, iron and levels of nutrition knowledge both before and after the intervention at 0.05 significance levels. Cross-tabulation was used to determine the changes in occupations of respondents, crop diversity, nutrition knowledge and dietary diversity in the households after intervention. The results are presented in tables and graphs below.

Research findings

Knowledge and attitudes

The farmers' knowledge, attitudes and skills depended upon the training they received. Knowledge in certain production techniques such as crop protection and rotation led to high levels of production (Mougeot, 2000). The project participants were offered agricultural and nutrition knowledge.

Agricultural knowledge

The results of the baseline survey revealed that none of the 300 respondents had received any kind of organized agricultural training. Because of their lack of knowledge, a training programme on organic farming was developed covering soil preparation, crop management, harvesting techniques and post-harvest handling. Increases in agricultural knowledge was observed through improved farming practices.

Nutrition knowledge

The results of the baseline survey showed that of the 300 respondents, 49.7 per cent had no knowledge of macronutrients (carbohydrates and proteins) or micronutrients (vitamin A and iron). Approximately 30 per cent knew what carbohydrates, proteins, vitamin A and iron were, and could give their sources. However, out of the 30 respondents selected for intervention, 15 (50 per cent) had no idea about any of the macro- or micronutrients. After training, of these 30 respondents who were taught about carbohydrates, proteins, vitamin A and iron and their sources, nine (20 per cent) still showed no knowledge of them. After re-examining respondents' educational levels, it was discovered they had never gone to school and, therefore, they had a limited ability to recall the topic of the training session.

Attitudes towards urban farming

The respondents were given statements that represented attitudes towards UA and were asked whether they agreed or disagreed with them. In the baseline survey, 97 per cent expressed positive attitudes. All 30 households both before and after intervention had positive attitudes towards urban farming and all thought it represented a good thing. They recommended their neighbours should consider adopting it as it enhanced household food supply. Also, they noted that the government should support urban farming. Therefore, the research showed that stakeholders in the area of UPAL should collaborate with the urban and peri-urban residents to push UPAL into the national policy agenda.

Farming Practices

Household food production systems

According to the results of the baseline survey, there was either no farming, or else little or unorganized farming being done. Focus-group discussions revealed that residents could not grow crops on their plots because livestock destroyed them as they could not afford fencing materials. The most vulnerable households selected for interventions were provided with fencing materials and were only expected to provide labour for fencing, land preparation, planting and managing the crops.

The 30 households chosen for intervention had neither crops nor small livestock. After intervention, identified food-production systems included subsistence farming systems and market-oriented crop production, where 46 per cent of the intervention households practised the latter type of farming. All 30 households produced crops and reared small livestock – such livestock were chosen because they required low inputs in terms of space and feeding. Moreover, the Nairobi City Council by-laws allow rearing of small livestock unless someone complains of a nuisance. In fact, during the intervention process, the 30 households received cockerels and chickens or ducks, which the project team delivered. In addition, the National Livestock Extension

Programme (NALEP) has also been introduced in Nairobi province (Z. Ishani and D. Lamba, unpublished work, 2007).

FOOD SECURITY

Dietary diversity

Out of the 300 households that were in the baseline survey, 57 per cent were in the medium group that had consumed 4–6 groups of different foods in the past 24 hours (Table 1.1). Similarly, out of the 150 households in the evaluation survey, 54.7 per cent were also in this medium group.

Among the 30 households in the intervention, 30 per cent had low dietary diversity of 1–3 groups before intervention, but after intervention shifted to the medium diversity of 4–6 food groups. Those 70 per cent with medium diversity before intervention increased to 86.7 per cent after intervention. Also post-intervention, 10 per cent of households moved from medium diversity to high diversity where 7–9 groups were consumed. These results are associated with increased crop and dietary diversity due to interventions. High levels of dietary diversity ensure adequate intakes of essential nutrients such as vitamins and minerals, as well as increased nutrient density (Hoddinott and Yohannes, 2002).

Nutrient intakes by children <5 years

Improvements were noted in both caloric and protein intakes of the children belonging to the two groups, as presented in Table 1.2. These improvements could be associated with increased amounts of both food and income from the sale of the surplus produce.

There was also improvement in both vitamin A and iron intakes, as seen in Table 1.2. This was probably due to increased production and higher consumption of fruits. Consumption of vegetables and livestock products increases vitamin A and iron intakes. Research demonstrated that children in

Table 1.1 *Dietary diversity based on RDA before and after intervention*

| Dietary diversity | Baseline | | | | Evaluation | | | |
| | 300 assessed | | 30 vulnerable | | 150 evaluated | | 30 intervened | |
	N	%	N	%	N	%	N	%
1–3 low	122	40.7	9	30.0	67	44.6	1	3.3
4–6 medium	171	57.0	21	70.0	82	54.7	26	86.7
7–9 high	7	2.3	–	–	1	0.7	3	10.0
Total	300	100	30	100	150	100	30	100

Table 1.2 *Nutrient intakes based on RDA before and after intervention*

		Calories		
	Before		*After*	
Below RDA	23	76.7%	20	66.7%
Above RDA	7	23.3%	10	33.3%
		Proteins		
	Before		*After*	
Below RDA	17	56.7%	12	40.0%
Above RDA	13	43.3%	18	60.0%
		Vitamin A		
	Before		*After*	
Below RDA	22	73.3%	18	60.0%
Above RDA	8	26.7%	12	40.0%
		Iron		
	Before		*After*	
Below RDA	29	96.7%	25	83.3%
Above RDA	1	3.3%	5	16.7%
Total	30	100%	30	100%

the urban households with self-provisioned food showed better health than do those children who do not have such access (Mougeot, 2006). Prior to this research, better health was discovered to be linked to improved nutrition brought about by increased dietary diversity and its accompanying increased nutrient intake.

Nutrition status of the 300 children

Of the 300 children observed during the baseline survey, 62 per cent were stunted, 53.7 per cent were underweight and 31 per cent were wasted (the children were classified into one category at a time and overlapping of conditions were not established). The children with high levels of wasting were included in the intervention.

The nutrition status of children in the 30 households before and after intervention showed mixed results. However, there was a general improvement in their health as shown by reduced numbers both of children who were severely and moderately stunted, as well as those who were underweight (Figure 1.2).

Those children who were mildly wasted increased from 16.7 per cent to 46.7 per cent after intervention. Those who had normal weight decreased from 73.3 per cent to 43.3 per cent after the intervention.

In general, the above results do indicate changes. However, the results cannot necessarily be directly linked to the impact of the project because one

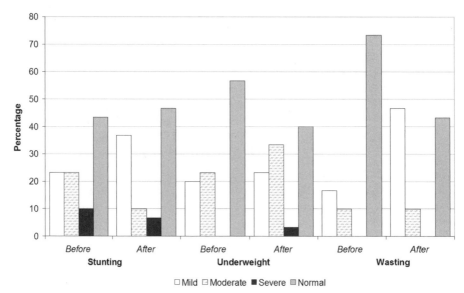

Figure 1.2 *Nutrition status of children from households involved in the intervention*

year is a short time for the project's activities to influence the children's nutrition status. Although adequate availability of food at the household level is necessary in order to achieve nutrition security, it is not sufficient. Other key contributors to good nutrition are adequate health, childcare, access to clean water and sanitation (FSAU, 2003).

RELATIONSHIPS BETWEEN VARIABLES

Between crop diversity and dietary diversity

A positive and significant relationship was found to exist between crop diversity and dietary diversity ($r = 0.123$ and $P = 0.03$), which shows that dietary diversity enhanced household nutrition and food security. Therefore, there is a need to promote production of a diversity of crops because it diversifies the households' diets, thereby enhancing household food security.

Positive relationships existed between the levels of nutrition knowledge and intakes of calories, proteins, vitamin A and iron, as shown in Table 1.3. This may have been due to availability of vegetables.

Improvement in the household food security after intervention

There was a significant difference in iron intake before and after intervention $t = 2.715$ and $P = 0.009$ (Table 1.4). This could be attributed to increased consumption of animal proteins and dark green leafy vegetables.

Table 1.3 *Nutrition knowledge and nutrient intakes*

Variables	R	Significant value P
Nutrition knowledge and calories	0.040	0.49
Nutrition knowledge and proteins	0.055	0.34
Nutrition knowledge and vitamin A	0.123	0.03*
Nutrition knowledge and iron	0.052	0.37

*Significant at 0.05 confidence level.

Table 1.4 *t-test for differences after intervention*

Variable	t-value	Mean difference	Significant value P
Caloric intake	1.636	0.567	0.107
Protein intake	1.504	0.267	0.138
Vitamin A intake	0.898	0.200	0.373
Iron intake	2.715	0.400	0.009*
Nutrition knowledge	2.843	0.967	0.060

*Significant at 0.05 confidence level.

There were insignificant differences found in the intakes of calories, proteins, vitamin A and nutrition knowledge before and after intervention.

CHANGES IN AGRICULTURE AND LIVESTOCK EXTENSION SERVICES

The staff of Ministry of Agriculture and Ministry of Livestock and Fisheries Development became interested in the activities of the project and introduced the multi-storey gardening (use of sacks to plant crops) in other households. Moreover, strong cooperation between the residents of Kamae and the agricultural and livestock extension officers at the divisional level has continued. This is a positive move towards pushing UPAL into the agriculture and livestock national policy agenda.

Other benefits associated with the project

During the focus-group discussions, the problem of access to clean and safe water came out strongly. The participants came together and formed a community-based organization under the guidance of the project team. They each decided to contribute approximately US$58 to purchase water pipes and water meters, while 15 participants connected to piped water and the remaining 15 were due for connection at a later date. Many non-participants have also managed to access the same water through the efforts of World

Vision's Kahawa Area Development Programme. Non-participants developed an interest in interventions: they fenced off their plots and planted crops.

CONCLUSION AND RECOMMENDATIONS

The findings of the study are very significant because NGOs and other interested individuals in Kamae saw it as a worthy cause and assisted community members in other innovations, especially in the provision of clean and safe water. Also, gardens were introduced into this community. Moreover, adjacent communities asked to be similarly trained and as a result of this, World Vision (Kahawa Area Development Programme) sponsored a six-day training programme in Soweto and also in Ngomongo, in Kasarani, Nairobi. Through the Ministry of Agriculture and Ministry of Livestock and Fisheries Development, the government should implement more training to all willing urban dwellers.

Fifty per cent of the households that participated in the interventions produced surplus produce they could sell. Thus, they became both farmers and business people. Thirty-six per cent in this same group were involved in farming as well as being engaged as casual labourers in the neighbouring coffee plantations. Only 13.4 per cent continued to be involved in farming without any other economic activity. Diversification of economic activities broadened the households' financial base or capacity and, as a result, access to household food security was enhanced. However, there is a need to sensitize urban dwellers on the importance of diversifying their economic activities for their overall development, which would include farming.

Agricultural knowledge increased significantly, as evidenced in the improved farming practices and diversity of crops which became accessible to households. Organic farming was the preferred system of crop production because it does not require expensive inputs; hence, it is affordable to people of all socio-economic classes. Those households participating in the intervention phase experienced an improvement in their food security and this also indicated an increase in dietary diversity due to improved caloric, protein, vitamin A and iron intakes. Nutrition education was essential so participants could make informed choices in the selection of food for their household's members and also so as to properly utilize available food.

Therefore, the research showed that when agricultural programmes are combined with well-designed nutrition education, significant changes in the participants' consumption behaviour can occur. Also, when there is an increase in nutrition knowledge there is a related increase in intakes of important nutrients, particularly vitamins and minerals. Increase in consumption of micronutrient-rich food is notable when production systems are diversified to include intercropping and rearing of small livestock.

The nutrition status of children during the baseline survey showed very high levels of stunting, which meant that they had experienced long-term deprivation of sufficient food and probably disease. However, the results of the intervention showed mixed responses, proving that nutrition education needs to be included in such interventions. Nutrition knowledge should be given to community members because it is absolutely essential for making informed choices in the selection and preparation of food for the household members. It is important to educate mothers on how to provide adequate nutrition and healthcare so as to improve the nutrition and health of children (especially those in their early childhood) to prevent stunting.

ACKNOWLEDGEMENTS

I want to acknowledge everyone who dedicated their time, effort and input to ensure the success of this project. I thank my supervisors Dr Elizabeth Kuria and Dr Margaret Keraka of Kenyatta University for their committed guidance and support.

Sincere thanks, also, to project-team members Ruth Nyansera, Jane Kaberia and Paul Okuta from the Ministry of Agriculture and Ministry of Livestock and Fisheries Development for their technical support. I also acknowledge the support of Gideon Ndung'u, my research assistant, Elizabeth Wambui and Josephat Wahiu, my field assistants, as well as to Onesmus Muinde and Robert Ofwago for technical advice in the interpretation of data. Lastly and most importantly, I sincerely appreciate the financial support of the AGROPOLIS granting programme of the Canadian-based International Development Research Centre (IDRC), which enabled me to complete the project successfully.

NOTES

1   Eunice Wambui Njogu, Ph.D. (Foods, Nutrition and Dietetics), PO Box 62697, Nairobi 00200, Kenya; email: koieunnice@yahoo.ca
2   In Fisher's formula, n was determined as follows: $n = Z^2 pq$ divided by d^2 where n = the desired sample size when the population is more than 10,000; Z = the standard normal deviation at the required confidence level; p = the proportion in the target population estimated to have characteristics being measured; $q = 1 - p$; and d = the level of statistical significance set. The calculation was: $n = (1.96)^2$ (0.50) (0.050) divided by $(0.05)^2 = 384.00$.

REFERENCES

Bonnard, P. (2001) *Improving the Nutrition impacts of Agriculture Interventions: Strategy and Policy Brief*, Food and Nutrition Technical Assistance Project (FANTA) and Academy for Education Development (AED), Washington DC
EpiInfo (2000) Anthropometric analysis software. US National Centre for Health Statistic NCHS and World Health Organization (software accessed in 2000; http://www.cdc.gov/epiinfo/)
Fisher, A. A., Lang, J. E., Stoeckel, J. E. and Townsend, J. W. (1998) *Handbook for Family Planning Operations Research Design*, 2nd edition, Population Council Nairobi, Nairobi

Foeken, D. and Mwangi, A. M. (2000) 'Increasing food security through urban farming in Nairobi', in Bakker, N., Dubbeling, M., Gundell, S., Sabel-Koschella, U. and de Zeeuw, H. (eds) *Growing Cities, Growing Food Urban Agriculture on the Policy Agenda: A Reader on Urban Agriculture*, Deutsche Stiftung fur Internationale Entwicklung (DSE), Feldafing, Germany

Food Security Analysis Unit for Somalia (FSAU) (2003) *Nutrition: A Guide to Data Collection, Analysis, Interpretation and Use*, Food Security Analysis Unit for Somalia, Nairobi, Kenya

Government of Kenya (2004) *Strategy for Revitalizing Agriculture (2004–2014)*, Ministry of Agriculture and the Ministry of Livestock and Fisheries Development, Nairobi Government Printer, Nairobi

Hoddinott, J. and Yohannes, Y. (2002) *Dietary Diversity as a Household Food Security Indicator*, Food and Nutrition Technical Assistance Project (FANTA) and Academy for Education Development (AED), Washington DC

Iannotti, L. and Gillespie, S. (2002) *Successful Community Nutrition Programming Lessons from Kenya, Tanzania and Uganda*, LINKAGES, Academy for Education Development (AED), Washington, DC

Mougeot, J. A. (2000) 'Urban agriculture: Definition, presence, potentials and risks', in Bakker, N., Dubbeling, M., Gundell, S., Sabel-Koschella, U. and de Zeeuw, H. (eds) *Growing Cities, Growing Food Urban Agriculture on the Policy Agenda: A reader on Urban Agriculture*, Deutsche Stiftung fur Internationale Entwicklung (DSE), Germany

Mougeot, J. A. (2006). *Growing better Cities, Urban Agriculture for Sustainable Development*, International Development Research Centre, Ottawa, ON, Canada

Nutrisurvey (2004) Nutrient analysis computer software, University of Indonesia, SEAMEO-TROPMEN (software accessed in 2005 http://www.nutrisurvey.de)

World Health Organization (WHO) (1995) *Uses and Interpretation of Anthropometry, Expert Committee on Nutrition and Physical Status*, WHO, Geneva, Switzerland

2

Urban Compost: A Socio-economic and Agronomic Evaluation in Kumasi, Ghana

Nikita S. Eriksen-Hamel and George Danso[1]

INTRODUCTION

Uncollected and poorly managed solid and liquid wastes are a health and environmental hazard, especially to the urban poor who live near informal, and often illegal, waste dumps. The lack of facilities to collect, transport and treat municipal, agricultural and industrial wastes poses a major challenge to the rapidly expanding cities of West Africa. To address this, new and innovative methods of collection, transport, processing and storage need to be identified. Composting of municipal wastes and using the compost for agricultural purposes is a quickly growing and viable option to managing urban wastes in both the developed and developing world (Drechsel and Kunze, 2001). Composting provides the environmental benefit of diverting waste from landfill sites, and health benefits by reducing the survival and spread of pathogens in wastes. Furthermore, the end product is a valuable soil fertilizer. The use of urban composts in urban and peri-urban agriculture (UA and UPA) effectively closes the nutrient cycle in urban areas and reduces nutrient losses to the environment (Drechsel and Kunze, 2001).

Kumasi is the second largest city in Ghana, with a population of 1,017,000 and a growth rate of 3.1 per cent (Ghana Statistical Service, 2002). It has a semi-humid tropical climate, with average annual rainfall of about 1500 mm. Its central location in the country and relatively well-developed infrastructure make it a major market and distribution centre for Ghana and other West African countries. Within the peri-urban area, which covers an area of approximately 40 km radius around the city centre (Blake and Kasanga, 1997), are many agri-food industries such as breweries, saw mills and commercial poultry farms, which produce large quantities of organic wastes. The daily domestic waste generation in Kumasi is about 610 tonnes. The two main markets generate an additional 250 tonnes per day, most of which is organic; thus, the total city-wide generation is about 860 tonnes per day (KMA-WMD, 2000). Farming and trading are the main income-generating activities in Kumasi. Commercial crop production is dominated by vegetable horticulture and staple crops such as plantain, maize and cassava. In partially waterlogged valley bottoms, sugarcane and taro are also grown.

Kumasi is an ideal city to promote composting as a waste management strategy due to the large amount of organic wastes that are generated, the high biodegradable organic fraction of solid waste ranging between 65 and 75 per cent and the extensive agriculture within its peri-urban boundaries (Salifu, 2001). A pilot composting plant was established in the Buobai suburb of Kumasi in 2001 by the IWMI and Kumasi Metropolitan Assembly (KMA). Different composts were produced from market wastes, household solid wastes and dewatered human sewage from septic and public toilets. During the inception of this composting project, the stakeholders wished to know the farmers' perceptions and demand for compost in Kumasi. At the onset of compost production, its agronomic potential and benefits to urban and peri-urban farmers had to be evaluated. This paper presents results of two composting studies.

OBJECTIVES

The objectives of this project were to conduct a social, economic and agronomic evaluation of compost made from urban organic solid wastes and fecal sludge, as well as an evaluation of its potential benefits in urban agricultural systems of Kumasi.

The social and economic evaluations were conducted through individual interviews and focus group discussions with relevant stakeholders. The objectives of these interviews were to: examine and document the farmers' perceptions, knowledge and socio-cultural acceptability of the compost; estimate the farmers' WTP for the compost compared with other common soil inputs; determine the socio-economic factors affecting farmers' WTP; and determine the demand for compost by farmers. Agronomic benefits of the compost were evaluated in seed germination and phytopathology experiments in the screenhouse and field trials. The objectives of these experiments were to: determine the success of germination and growth of vegetables to increasing rates of compost; determine whether the compost suppresses root knot nematodes, *Fusarium* wilt and *Sclerotium rolfsii* in lettuce and tomato plants; compare the fertilizer value of the compost with that of poultry manure; and determine the residual nutrient effects of compost applications on lettuce yield and soil fertility.

METHODOLOGY

Data collection from field interviews and focus group discussions

Interviews and focus group discussions with stakeholders, farmers and potential customers used a partially pre-coded questionnaire to elicit WTP and determine attributes of farmers' perception and farming practices. Additional data were collected from government institutions, IWMI urban and peri-urban agriculture (UPA) library, Ghana office of the Food and Agriculture Organisation of the United Nations (FAO), Ghana Organic Agriculture Network (GOAN) library and farmer associations. A total of 200 individual interviews were conducted with farmers from different urban farming systems in and around Kumasi. Few women were involved in agriculture in Kumasi; therefore, specific attempts were made to interview as many women as possible to elicit their opinions. The contingent valuation method (CVM) was used during interviews with farmers to estimate the price they would pay for compost based on their knowledge of similar products such as poultry manure or black soil. The CVM approach involved surveying individuals to determine whether and how much they are willing to pay for compost under different hypothetical scenarios (Whittington et al, 1990; Asenso-Okyere et al, 1997; Alberni and Cooper, 2000). Farmers were asked whether they were willing to pay for compost as a soil improver and how much they were willing to pay for a 50-kg bag. A bidding game was introduced in situations where farmers found it difficult to understand and provide their bids. This bidding game helped to determine the zero WTP, and the mean and maximum amount the farmers would pay for the compost (Field, 1994; Hanley et al, 1997; Akpalu, 2000; Nugent, 2001).

Further questions focused on socio-economic characteristics, income, experience with compost and perception of compost quality. Since the dependent variable was a dichotomy (yes – 1 or no – 0), it was deemed appropriate to use a probit model instead of traditional regression models. The probit model was used to explain the factors that could affect farmers' WTP for compost (Kennedy, 1990) (Table 2.1). The probit model is defined as:

$$\Pr(y = 1 | x) = \Phi(xb)$$

where Φ is the standard cumulative normal probability distribution and xb is the probit score and takes care of the explanatory variables (Pindyck and Rubinfeld, 1983). The parameters in the probit model were estimated by maximum likelihood methods (Pindyck and Rubinfeld, 1983; Hailu, 1990).

Compost production and analysis

The compost plant produced four types of compost with differing ratios of household waste, market waste and dewatered fecal sludge. The first compost

Table 2.1 *Chemical characteristics of soil fertility amendments*

Treatment	TKN (%)	Org C (%)	Org M (%)	P (%)	K (%)	EC (ms/cm)	pH
Compost	1.06	6.75	11.64	1.84	3.22	n.a.	7.3
Potting mixture	0.06	n.a.	0.75	n.a.	n.a.	0.31	7.0
Poultry manure	1.14	4.91	8.50	n.a.	n.a.	7.86	7.4

EC, electrical conductivity; TKN, total Kjeldahl nitrogen.

produced and evaluated in field trials was made from a 3:1 ratio of market waste and dewatered fecal sludge. Organic market wastes were obtained from the main waste bins of the Kumasi Central Market. The sludge was obtained from public toilet septic tanks and was dewatered on a drying bed at the compost plant. Details of the dewatering facilities and compost production are described in the IWMI report (2003). The compost, poultry manure and test soils were analysed using established methods at the Soil Research Institute (SRI) in Kumasi and are presented in Table 2.1.

Germination experiment

The eight common vegetables used in the germination experiment were tomato (*Lycopersicon esculentum*), sweet pepper (*Capsicum annuum*), okra (*Hibiscus sabdariffa/esculentus*), carrot (*Daucus carota*), garden eggplant (*Solanum melongena*), lettuce (*Lactuca sativa*), cabbage (*Brassica oleracea capitata alba*) and spring onion (*Allium cepa*). Thirty seeds of each vegetable were planted into trays filled with one of five soil treatments. The soil treatments were a mixture of potting soil and fresh compost at 0, 12.5, 25, 50 and 100 per cent. The seeds were loosely placed on the surface and lightly covered with soil. Okra seeds, being the largest seeds used, were soaked in water overnight (16 h) before planting. Seedling trays were placed in a screenhouse for three weeks, where temperatures ranged between 25 and 30°C, and watered daily to maintain sufficient moisture.

Germination was counted every day for 15 days and was considered successful when the seed radicle protruded from the surface of the soil. The average time to emergence (ATE) was calculated as:

$$\text{ATE} = \frac{[\sum(\text{number of germinated seeds on } i\text{th day} \times i\text{th day})]}{(\sum \text{ germinated seeds on last day})}$$

After three weeks seedlings were harvested by uprooting and washing the root soil. The length of the primary taproot, shoot and fresh weights were also measured.

Nematode suppression experiment

A factorial experiment was designed with two vegetables (lettuce and tomato plants), six soil treatments and two nematode treatments (with and without nematodes present) and replicated four times. The six soil treatments were prepared with potting mixture and the following fresh-weight percentages of compost: 0, 6.25, 12.5, 25, 50 and 100 per cent. For each vegetable, two three-week-old seedlings were transplanted into a half-litre pot containing a different soil treatment. Into half of the pots, a 1-ml aliquot of nematode egg suspension, prepared using established methods, was injected into the soil 1 cm from the seedlings. Pots were kept in a screenhouse for six weeks, with temperatures ranging between 25 and 30°C, and watered every two days to maintain sufficient moisture. At harvest, plants were uprooted and shoot, root lengths and fresh weights determined. The incidence and severity of nematode root-galling was rated using a five-point scale.

Field fertilizer experiment

Two field experiments were conducted with the farmers of the Gyenyasi Farmers Cooperative. This cooperative has about 16 farmers, who produce lettuce, cabbage and spring onions in a 2-ha plot. Fourteen raised lettuce beds belonging to three farmers were divided into 30 plots, with sizes ranging from 7.5 to 12 m². The fertilizer treatments included the compost and poultry manure applied at rates of 1, 2 and 4 kg m⁻² on each crop for three crops, compost applied at 2.5, 5 and 10 kg m⁻² on the first crop, and a control with no compost applied.

The first experiment compared the compost and poultry manure applications on three crops. Since farmers evaluate the impact fertilizers based on how much they have to carry into the field, the two fertilizers were compared on a weight basis and not a nutrient basis. The second experiment compared lettuce yields over three harvests from a single application of compost to the first crop. Lettuce plants were watered daily with approximately 6 l m⁻² of irrigation water containing about 28 mg N l⁻¹. The equivalent fertilization was about 50 kg N ha⁻¹ per month. Hand weeding was done on the same day each week and no pesticides were applied on the first crop. An outbreak of septoria leaf spot (*Septoria lactucae* Pass.) in the second and third crops required the use of Dithane fungicide. Composite soil samples were taken for chemical analyses and bulk density determined from each plot at the beginning of the experiment and after each cropping season. The fresh yield of all lettuce plants within a 1-m² grid was determined four weeks after transplanting.

Statistical analysis

Descriptive and inferential statistical tools were used to analyse interview data using SPSS software. The data from the CVM survey were analysed with a probit model. Correlations and means comparisons were conducted for the different explanatory variables to avoid multi-colinearity in the probit model. Due to difficulties in obtaining sufficient materials and labour, the germination experiment was not replicated and only one experimental unit per treatment was available. The unreplicated data were analysed by regression analysis. The effects of different compost treatments on nematode galling and yield were evaluated using ANOVA procedures. All the treatment comparisons were carried out at the $P = 0.05$ significance level.

RESEARCH FINDINGS

Social perceptions of compost made from urban organic wastes

Generally, the farmers had a good perception of the quality of compost regardless of how much experience they had using it. There was consensus that compost improved soil quality and increased crop yield. The major reasons given for this were: compost resembles 'black soil', which is known to farmers as a good soil input; high quantities of 'night soil' mixed with compost enriched the biofertilizer; they perceived long-term effects on the soil; and they associated compost use with benefits related to waste reduction and urban sanitation. About 42 per cent of the farmers thought that handling compost did pose the same risks associated with the handling of chemical fertilizers.

The 20 per cent of farmers who had no experience with compost perceived it to be ineffective, largely due to cultural beliefs or health concerns of the night-soil component of the compost. They believed that handling compost with night soil would lead to spreading of diseases such as HIV, typhoid and cholera. Furthermore, these farmers assumed that consumers would avoid crops on which compost had been used for fear of infections. Other reasons for being sceptical about the compost were the high labour requirements for compost production (57 per cent); insufficient amount of raw materials to produce compost (28 per cent); and the small market for organic products in the Kumasi region (15 per cent). A similar survey in Kumasi found that farmers had the same perceptions and reasons for accepting compost made from urban wastes (Warburton and Sarfo-Mensah, 1998).

The majority of farmers indicated that compost would be moderately (51 per cent) or very difficult (10 per cent) to apply to soils compared with chemical fertilizers, which could easily be applied with irrigation water. Poultry manure was smelly, cloddy and inconvenient to apply when wet, whereas compost could be spread more easily using rakes. Many farmers perceived compost to be expensive (48 per cent), but were willing to handle it (42 per cent) and believed it could control the incidence of pest and diseases (42 per cent). The cost of establishing a compost plant and the increased cost of transporting the material were the main economic concerns of farmers. Few were unwilling to handle the compost (6 per cent) or believed it could not control the incidence of pests and diseases (2 per cent). However, the majority of farmers were indifferent about handling compost (52 per cent) or had no opinion about its ability to control pests and diseases (56 per cent).

Their claim was that, since they had not used it before, they could not judge its effectiveness at controlling pests and diseases.

Farmers' willingness to pay for compost

There was a positive correlation between how farmers perceived compost and their WTP for it. About 70 per cent of them expressed positive WTP, while 30 per cent were unwilling to pay even though they perceived it as a good soil input. Farmers were unwilling to pay for it largely for economic reasons. They were generally satisfied with the high returns achieved from their current soil input and did not see a need to replace or try new ones. Many suggested that it should be the government's responsibility to subsidize or supply inputs free to farmers.

Comparing WTP bids across different farming systems in Kumasi revealed that urban vegetable farmers expressed the lowest mean WTP (US$0.10), although their systems were profitable, with annual revenues ranging between US$600 and $1000 per hectare. Peri-urban vegetable farmers proposed the highest mean WTP (US$3; Table 2.2). The lower bid from urban vegetable farmers may have been influenced by the cheap and readily available poultry manure in urban farming systems.

A previous study in Kumasi found that more than 90 per cent of the urban vegetable farmers preferred poultry manure because of the low price (US$0.01 per 10 kg) and availability (Drechsel and Kunze, 2001). On the other hand, the peri-urban vegetable farmers used expensive chemical fertilizers, which cost about US$2.80 per 10 kg. Although the nutrient content was greater in chemical fertilizers, the cost of nitrogen was still cheaper for the poultry manure than the chemical fertilizer.

The poultry industry in Kumasi may have the potential to provide large amounts of bio-fertilizer. There are about 300 registered commercial poultry farms in and around Kumasi (Kindness, 1999), and manure production by the largest farm alone is about 50 tonnes per week (Drechsel, 1996).

Decision variables for farmers' WTP

Farmers' experience of the use of compost had significant influence on the WTP for compost, as did gender and household dependency. However, their

Table 2.2 *Farmers' willingness-to-pay for 50 kg of compost*

Farming system	Mean WTP US$ (Std. Dev.)
Vegetable farming, urban	0.10 (0.10)
Vegetable farming, peri-urban	3.00 (1.10)
Staple crops, urban	2.00 (1.60)
Staple crops, peri-urban	2.70 (1.00)
Backyard, urban	1.40 (0.10)
Ornamentals	0.60 (0.40)

Table 2.3 *Probit model of explanatory variables of farmers' WTP for compost in Kumasi*

Independent variables	Regression coeff.	Standard error	t-statistics[1]
Perception	0.566	0.077	7.3**
Location	0.062	0.051	1.2
Gender	−0.043	0.044	−0.98
Education	0.178	0.067	2.1**
Age	0.043	0.022	2.0*
Income	0.000	0.000	−2.2*
Household dependency	−0.078	0.043	−1.8
Compost experience	−0.119	0.105	−1.1
Soil inputs	−0.127	0.043	−2.9**

[1]Coefficient is significant at 5 per cent (*) and 1 per cent (**).

perceptions as related to compost, household income, level of education and age were all significant variables that influenced WTP (Table 2.3).

As expected, a positive perception of compost implied a higher probability of WTP for the compost. Similarly, farmers with higher income were willing to pay more than poorer farmers. Farmers who could read and write have a higher probability of paying for compost than those who could not read or write. This was most likely because more educated farmers were better informed about innovations and better understood the advantages and disadvantages of compost. Older farmers had a lower WTP than younger farmers, possibly due to older farmers not valuing the need to invest in the land they were using, whereas younger farmers were more motivated to invest in their land to realize long-term benefits. The present soil input the farmer was using had a significant impact on his or her WTP for compost. This was primarily because farmers used inexpensive soil fertility improvement inputs which were familiar to them. The farmers who were not interested in applying compost to their crops had less security on the plots they farmed and long-term investment in soil quality was not a priority to them.

Demand for compost in Kumasi

Demand for compost in Kumasi was estimated through WTP bids by farmers for both the subsidized and unsubsidized compost plants. Compost produced from Kumasi was valued at US$5 per 50 kg bag and its demand was 940 tonnes per year. Peri-urban farmers were willing to pay US$4 per 50 kg bag, which created a difference of US$2 with the prevailing market prices. This meant that, by sustaining demand for compost at this higher price, the government would have to subsidize the prices by US$2, which would lead to an explosion in demand estimated at 11,000 tonnes per year.

Although the compost plant is centrally located with respect to the urban farming areas of Kumasi, the distances between it and the farms are still

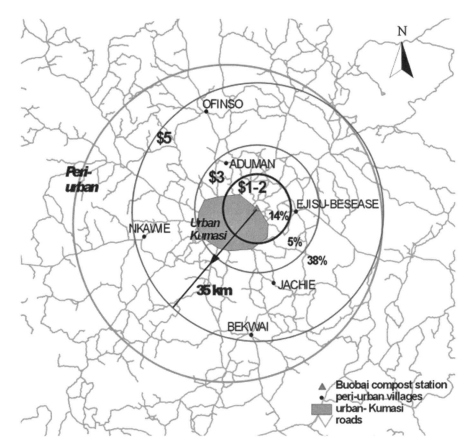

Figure 2.1 WTP *for compost at different distances from the compost station*

large, plus the road access to these farms is often poor. Even with subsidized compost production, high transportation costs make it difficult for those living beyond a 35-km radius of the compost plant to afford to buy compost (Figure 2.1). Since peri-urban farms are located beyond a 40-km radius of the city (Adam et al, 1999), it appears unlikely that composting of urban waste in Kumasi would be a realistic option for recycling nutrients from urban areas to peri-urban and rural areas.

Germination experiment

The germination percentage was consistent for all vegetables and ranged from 70 to 100 per cent except for okra and garden eggplant, whose seeds were infested by red ants. The germination of most vegetables was unaffected by the compost application, with the exception of carrot and onion. Germination of carrot ($P = 0.050$, $R^2 = 0.55$) and onion ($P = 0.050$, $R^2 = 0.61$) decreased as the rates of compost application increased. Seedling survival of most vegetables was similarly unaffected by compost rates, with the exception of carrot ($P = 0.048$, $R^2 = 0.78$) and onion ($P = 0.060$, $R^2 = 0.74$). The average time of germination increased with increasing compost rates for tomato ($P = 0.031$, $R^2 = 0.83$), carrot ($P = 0.013$, $R^2 = 0.86$), lettuce ($P = 0.012$, $R^2 = 0.91$) and okra ($P = 0.001$, $R^2 = 0.98$). For these vegetables, the average time for germination was 1–2 days longer at high compost rates (>50 per cent) than low rates (<6.25 per cent). Differences in shoot weight were not apparent after 21 days except for okra, which had the greatest shoot biomass at 6.25 per cent compost.

With the exception of carrot and onion plants, seed germination and seedling growth of most vegetables were unaffected by the phytotoxic compounds in the compost. Compost applications to carrot and onion nursery beds should be limited to low amounts or entirely avoided. The delay in germination of some plants is most likely due to low levels of salinity in the compost, which interfered with the imbibition phase of the seeds.

Nematode suppression experiment

Root galls were observed for the majority of plants inoculated with root knot nematodes. Tomato and lettuce plants without compost and those infected with nematodes were either dead or highly infected at the time of harvest. Root galling in tomato ($P = 0.001$, $R^2 = 0.70$) and lettuce plants ($P = 0.002$, $R^2 = 0.97$) decreased with increasing compost rates. A significant reduction in root galling from the 0 per cent compost treatment was observed in tomato plants at 6.25 per cent compost and in lettuce plants at 50 per cent compost. Phytotoxic effects of compost significantly reduced plant weight when compost rates exceeded 50 per cent in tomato and 25 per cent in lettuce. The optimum compost rate for greatest yield was between 6.5 per cent and 12 per cent for tomatoes and 12 per cent for lettuce. This research supports other findings which have shown that composted biowaste reduces the incidence of many diseases in a wide range of vegetables (Tuitert et al, 1998; Blok et al, 2002).

Field fertilizer experiments

Three farmers volunteered a limited number of lettuce beds to be used in the experiment. The resulting experimental design had each farmer managing a single replicate of each treatment on their lettuce beds. This design introduced the farmers as a new factor along with fertilizer type and application rate. To discount the effects of the farmers, we attempted to control the management of the lettuce beds. The date of planting and harvesting, quantities and method of fertilizing, watering and hand-weeding were the same for all of them. Nonetheless, differences between them were confirmed through statistical analysis ($P = 0.01$) and these were caused by differences in management history, location of lettuce beds and pest incidence among the three farmers' plots. Contrast analysis showed that poultry manure treatments had greater yield ($P = 0.004$) than compost treatments. The compost and poultry manure treatments with the same application rates

Table 2.4 *Yield of lettuce following application of the compost and poultry manure*

| Treatment | Rate | Time of application | Crop[1] | | | Mean of all crops $(kg\,m^{-2})$ |
			1st $(kg\,m^{-2})$	2nd $(kg\,m^{-2})$	3rd $(kg\,m^{-2})$	
Control	$0\,kg\,m^{-2}$	n.a.	0.9 b	1.0 b	0.8 cd	0.9 d
Compost	$1\,kg\,m^{-2}$	Every crop	1.4 ab	1.2 ab	1.3 bcd	1.3 cd
Compost	$2\,kg\,m^{-2}$	Every crop	1.2 ab	0.9 b	1.4 abcd	1.2 d
Compost	$4\,kg\,m^{-2}$	Every crop	2.4 a	1.7 ab	1.6 abc	1.9 abc
Poultry manure	$1\,kg\,m^{-2}$	Every crop	2.0 ab	1.7 ab	2.2 a	2.0 ab
Poultry manure	$2\,kg\,m^{-2}$	Every crop	1.8 ab	1.3 ab	1.8 ab	1.6 abc
Poultry manure	$4\,kg\,m^{-2}$	Every crop	2.4 a	2.3 a	2.0 ab	2.2 a
Compost	$2.5\,kg\,m^{-2}$	First crop	1.4 ab	1.0 b	0.7 d	1.1 d
Compost	$5\,kg\,m^{-2}$	First crop	1.7 ab	1.2 ab	0.7 d	1.2 d
Compost	$10\,kg\,m^{-2}$	First crop	2.2 ab	1.6 ab	0.7 d	1.5 bcd

[1]Values in each column for each crop and the mean of crops followed by similar letters are not significantly different than Tukey's HSD test ($P < 0.05$). Letters indicate treatment groupings that are statistically different from one another. Grouping rank decreases with alphabetical order (e.g. group 'a' is greater than group 'b').

were paired on the same bed to allow for direct visual comparisons. Although the only significant differences between paired treatments were at the lowest application rate ($1\,kg\,m^{-2}$) in the third crop (Table 2.4), all farmers agreed that lettuce plants were larger and greener in poultry manure treatments compared with compost treatments. Furthermore, the farmers' perceptions agreed with the statistical analysis, which showed that $4\,kg\,m^{-2}$ compost had greater lettuce yield ($P < 0.05$) compared with the control but only in the first crop. Marketable yields of $1.5\,kg\,m^{-2}$ were obtained from all poultry manure treatments and from the highest compost application ($4\,kg\,m^{-2}$) (Table 2.4). Lettuce yields increased linearly with increasing compost application in the first crop and marketable yields (about $1.5\,kg\,m^{-2}$) were achieved for the 50 and $100\,kg\,N\,m^{-2}$ treatments. In the second crop, marketable lettuce yields were attained only with addition of the $100\,kg\,N\,m^{-2}$ treatment.

Lettuce yields in treatments that only had a compost application on the first crop were not different from the control in each crop nor for the mean of all crops. Yields obtained from the control treatment remained the same for the crops in the three cropping seasons. Since no soil nutrients were added, it would have been expected that the continued extraction of nutrients in harvested lettuce would deplete the nutrients in the control plot to a level where no crop would have grown. This is because nitrogen mineralized from the organic matter in the sandy soil was unlikely to provide sufficient nutrients to explain these yields. However, irrigation water used in this study supplied about $50\,kg\,N\,ha^{-1}$ per month and acted as fertilizer to the control treatment.

CONCLUSIONS AND RECOMMENDATIONS

Farmers were aware of the potential benefits compost could give to improving soil quality, increasing yields and controlling the incidence of pests and diseases. However, there were many concerns regarding the potential costs of producing compost, the difficulty in using the bulky compost and the health risks associated with handling compost made from night soil.

WTP for compost was lowest for urban vegetable farmers (US$0.10 per 50 kg), who rely on readily accessible and cheap poultry manure, whereas peri-urban vegetable farmers were ready to pay US$3 per 50 kg.

Demand for compost in Kumasi was estimated at about 11,000 tonnes per year based on the assumption that subsidies were provided. But without subsidies actual demand was only 940 tonnes per year.

Adding compost increased growth and did not delay germination of a wide range of vegetables when applied at concentrations between 6.25 and 12.5 per cent, with the exception of carrot and onion. Nematode damage in lettuce and tomato plants was also suppressed, and poultry manure was the organic fertilizer of choice for the urban farmers in Kumasi.

In conclusion, we can infer from this study that large-scale compost production from urban organic waste and night soil in Kumasi is not an economically viable project without some kind of subsidy. Sorting and transportation costs for compost at this scale outweigh the economic benefits associated with the use of compost. There is a need to explore also the viability of community-based composting initiatives at the farm level, where sorting and transportation costs are minimal.

ACKNOWLEDGEMENTS

This research project benefited from the support and time of many individuals and organizations. The authors would like to thank the generous financial assistance of the International Development Research Centre (IDRC; AGROPOLIS programme), the IWMI, Wageningen University and Kwame Nkrumah University of Science and Technology (KNUST). The compost plant was funded by the IDRC, the French Foreign Ministry, the Swiss Federal Institute for Water and Sanitation in Developing Countries (SANDEC) and the KMA. We would like to acknowledge the support, help and guidance of Dr S. Fialor, Dr W. Blok, Prof. C. Quansah, Dr R. Awuah, Mr K. Osei, Dr P. Drechsel, Mrs K. Dapaah and all the technicians, labourers and farmers who helped us throughout this project. A special mention to the head of the composting project, Dr C. Olefunke, for her dedication to this project.

NOTE

1   Eriksen-Hamel, 34 Maple St. Apt. 31, Ste-Anne-de-Bellevue, Québec, Canada, H9X 2E6, tel: +1-514-457-7352/fax: +1-514-398-7990; email: nikita.eriksenhamel@gmail.com

REFERENCES

Adam, M. G., Sarfo-Mensah, P. and Warburton, H. (1999) 'Urban blight and farmers' plight: Peri-urban agriculture around Kumasi, Ghana', *The Land* 2(2), pp131–148

Akpalu, W. (2000) 'Willingness to pay for fish conservation and to reduce health risks associated with pollution of Fosu Lagoon, application of contingent valuation methods', *Proceedings of the Beijer Research Seminar*, Morocco, March 2000

Alberni, A. and Cooper, J. (2000) 'Application of the contingent valuation methods in developing countries', FAO Economic and Social Development Paper No 146, F, Rome, Italy

Asenso-Okyere, W. K., Osei-Akoto, I., Anum, A. and Appiah, E. N. (1997) 'Willingness to pay for health insurance in developing economy. A pilot study of the informal sector of Ghana using contingent valuation', *Health Policy*, 42, pp223–237

Blake, B. and Kasanga, K. (1997) *Kumasi Natural Resource Management Research Project*, Inception Report, Natural Resource Institute, University of Greenwich, UK

Blok, W. J., Coenen, T. C. M., Piji, A. S. and Termorshuizen, A. J. (2002) 'The Netherlands – suppressing disease in potting mixes with composted biowastes', *Biocycle*, 43, pp58–68

Drechsel, P. (1996) 'Applied research for peri-urban areas', *IBSRAM Newsletter*, 42, pp5–7

Drechsel, P. and Kunze, D. (eds) (2001) *Waste Composting for Urban and Peri-urban Agriculture – Closing the Rural–Urban Nutrient Cycle in Sub-Saharan Africa*, CABI Publishing, Wallingford, UK

Field, C. B. (1994) *Environmental Economics: An Introduction*, McGraw-Hill College, Maidenhead, UK

Ghana Statistical Service (2002) *Population Census of Ghana*, Ghana Statistical Service, Accra, Ghana

Hailu, Z. (1990) *The Adoption of Modern Farm Practices in African Agriculture. Empirical Evidence about the Impacts of Household Characteristics and Input Supply Systems in the Northern Region of Ghana*, Nyankpla Agricultural Research Report 7, CSI/GTZ, pp114–167

Hanley, N., Shogren, J. and White, B. (1997) *Environmental Economics in Theory and Practice*, Oxford University Press, Oxford, UK

Henn, P. (2000) 'User benefits of urban agriculture in Havana, Cuba: An application of the contingent valuation methods', MSc Thesis, McGill University, Canada

IWMI (2003) *Co-composting of Faecal Sludge and Solid Waste for Urban and Peri-urban Agriculture in Kumasi, Ghana*, Final Technical Report, International Water Management Institute (IWMI), Accra, Ghana

Kennedy, P. (1990) *A Guide to Econometrics*, 2nd edition, Basil Blackwell, Oxford, UK

Kindness, H. (1999) *Supply and demand for soil ameliorants in peri-urban Kumasi*, KNRMP R6799, NRI-DFID-KNUST Report, Kumasi, Ghana

KMA-WMD (Kumasi Metropolitan Assembly – Waste Management Department) (2000) *Yearly Report 2000*, Kumasi Metropolitan Assembly, Kumasi, Ghana

Nugent, R. A. (2001) 'Using economic analysis to measure the sustainability of urban and peri-urban agriculture: A comparison of cost-benefit and contingent valuation analyses', *Proceedings of the CIP-SUIPA and ETC-RUAF workshop on Appropriate Methodology in Urban Agriculture Research, Planning, Implementation and Evaluation*, Nairobi, Kenya

Pindyck, R. S. and Rubinfeld, D. L. (eds) (1983) *Econometric Models and Econometric Forecasts*, McGraw Hill, Tokyo, Japan

Salifu, L. (2001) 'An integrated waste management strategy for Kumasi', in Drechsel, P. and Dagmar, K. (eds) *Waste Composting for Urban and Peri-urban Agriculture: Closing the Rural–Urban Nutrient Cycle in Sub-Saharan Africa*, CABI Publishing, Wallingford, UK, pp112–115

Tuitert, G., Szczech, M. and Bollen, G. J. (1998) 'Suppression of *Rhizoctonia solani* in potting mixtures amended with compost made from organic household waste', *Phytopathology*, 88, pp764–773

Warburton, H. and Sarfo-Mensah, P. (1998) *The Use of Composted Urban Waste in Integrated Pest Management Systems to Control Pests and Pathogens in Peri-Urban Agriculture*, Technical Project Report C1045, DFID, RNRRS, and NRSP Peri-Urban Interface Program, Kumasi, Ghana

Whittington, D., Briscoe, J., Mu, X. and Barron, W. (1990) 'Estimating the willingness to pay for water services in developing countries: A case study of the use of contingent surveys in Southern Haiti', *Economic Development and Cultural Change*, 38, pp293–311

3

Urban Agriculture as a Livelihood Strategy in Lima, Peru

Luis Maldonado Villavicencio[1]

INTRODUCTION

This study was a component of the baseline of the project entitled 'Agricultores en la Ciudad' (Farmers in the City), carried out by the Urban Harvest Program hosted by the International Potato Center (CIP). This research used the neighbourhood of Carapongo, in the city of Lima, capital of Peru, as a case study. Lima has a population of more than seven million people and, like many Latin American cities, has a high concentration of urban poverty along with rampant and chaotic urban sprawl. At present, approximately 3.3 million poor people live in the city. Poverty in Peru has become a problem with a distinctly urban face. The explosive growth of Lima can be traced to 1940 when people migrated from rural areas to the capital in search of better living and working conditions. Growth occurred largely through the illegal occupation of both privately and publicly owned uncultivated lands.

Carapongo was selected as a case study because it represents a peripheral area of Lima where urban agriculture (UA) is commonplace. Here, one of the principal threats to urban producers is rampant urban sprawl. In addition, urban households involved in agriculture contend with a lack of recognition and understanding from policy makers. City authorities often still perceive agriculture as a rural activity conducted far away from the city and thus there is a marked lack of interest in the sector. In this context, an understanding of the strengths and limitations of UA becomes a crucial starting point for developing adequate political strategies.

This paper analyses UA as a household strategy which is used to confront urban poverty, and it examines institutional and political factors that block and/or support the strategy. Specifically, the paper identifies which assets households use in order to overcome urban poverty and how they engage with institutions and municipal authorities. Gender analysis between men and women in the use of household assets was also done.

The paper begins by explaining the methods used as well as the livelihoods approach. An analysis of access to assets by urban farmers is presented, followed by an overview of the dimensions of gender related to home-based agriculture. The paper then identifies the influence of UA in Carapongo and presents an analysis of the stakeholders and some of the conflicts between them. Finally, the study concludes with the main research results.

METHODOLOGY

Definition of the study area

Carapongo is located on the Rímac River approximately 15 km from the centre of metropolitan Lima (Figure 3.1). Here, many vegetables are produced by smallholder farms by a predominantly migrant population who come from rural areas. This locality was selected for several reasons: agriculture represents a way of life for more that half of its population (INEI, 2002); there is a strong interconnection and integration with the economic and ecological system of the urban centres; there is ample future potential due to the abundance of natural resources such as water and land; and producers can benefit from contacts with nearby organizations.

Methodology

In this study, several techniques and instruments of quantitative, qualitative and participatory research were used. Additionally, a review of literature, including a statistical overview, was carried out in order to prepare the theoretical framework.

In order to better understand the actors' various perceptions of UA, semi-structured, in-depth interviews were carried out. In total, 15 farmers and five local authority representatives were interviewed. Also, participatory workshops explored various subjects such as historical events that have affected the community in recent years. An example of this was the identification and prioritization of the principal historical problems affecting UA along with discussions and suggestions of alternative solutions – the matrix of alternatives.

Two discussion groups were conducted to collect information about gender roles, one comprised of men and the other of women. Other tools such as seasonal calendars, transect walks and problem analysis using drawing were also carried out for gender analysis.

For the survey, 125 households were randomly selected using a list of farmers. In each household, the person identified as the main individual responsible for the farm was interviewed. Interviews contained 110 questions and lasted approximately 45 minutes. All interviews were conducted by the author and a team of students during two months.

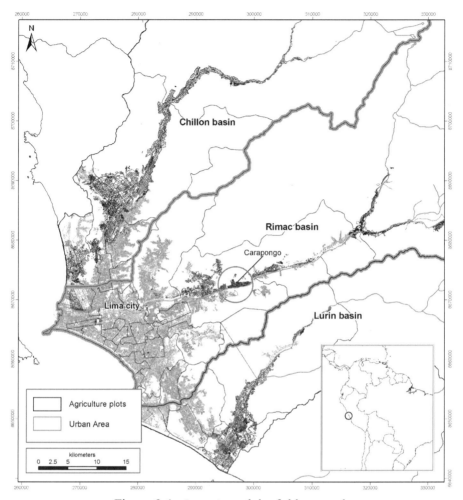

Figure 3.1 *Location of the field research*

The survey sought to identify the ways households identify assets. In addition to these activities, video images were recorded as a means of direct observation so as to validate and/or obtain basic information on the economic condition and level of services available in the area.

CONCEPTUAL FRAMEWORK

The livelihoods system framework

Chambers and Conway (1992) state:

> *A livelihood comprises the capabilities, assets (including both material and social resources) and activities required for a means*

of living. A livelihood is sustainable when it can cope with and recover from stresses and shocks, maintain or enhance its capabilities and assets, while not undermining the natural resource base.

The livelihoods framework was initially designed to improve the understanding of rural households, but it is now seen as a generic framework for use in urban as well as rural areas (Singh and Gilman, 1999; Martín et al, 2000; Sanderson, 2000). The framework does not attempt to provide an exact representation of reality, but it does provide a way of thinking about livelihoods designed to stimulate debate and reflection. The livelihoods framework views poor households as being dependent upon a diversity of strategies in order to face urban poverty (Figure 3.2). These strategies are based on a set of household 'assets': natural capital (land and water); financial capital; physical capital (houses, equipment, animals, seeds); human capital (in terms of both labour power and capacity, or skill); and social capital (networks of trust between different social groups). The deployment of assets also depends on external influences such as dealing with regulations, policies,

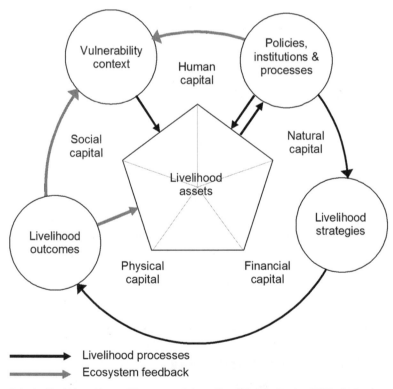

Adapted by Urban Harvest Program – International Potato Center (CIP), design by Thomas Zschoke from DFID, 2001, Sustainable Livelihoods Guidance Sheets.

Figure 3.2 *The sustainable livelihoods systems framework*

urban authorities and local marketing practices. The inability to adequately use and employ the various assets at their disposal can leave households vulnerable to economic, environmental, health and political stresses and shocks.

Various studies illustrate how UA is used as a strategy by the urban poor to generate income and provide fresh produce to urban consumers. The practice of UA is considered an important tool of the urban poor in order to contend with poverty, generate income and deal with insecurities such as procuring food (Rakodi, 1991; Maxwell, 1994; O'Reilly and Gordon, 1995; Antweiler, 2000; Arm-Klemesu, 2000; Nugent, 2000). Other work has analysed the crucial role played by women in UA (Rakodi, 1988; Mbiba, 1993; Maxwell, 1994; Mianda, 1996; Mudimu, 1996; UNDP, 1996; Hovorka, 1998; Mougeot, 2000).

This paper analyses UA as a livelihood strategy, using the following definition:

> *An industry located within (intra-urban) or on the fringe (peri-urban) of a town, a city or a metropolis, which grows and raises, processes and distributes a diversity of food and non-food products (re-)using largely human and material resources, products and services found in and around that urban area, and in turn supplying human and material resources, products and services largely to that urban area. (Mougeot, 2000)*

RESEARCH FINDINGS

Urban agriculture: The Carapongo case

In Peru, the importance of UA increased with the implementation of land reform during the 1970s. Prior to land reform, much of the agricultural land around Lima was occupied by large estates – haciendas – which primarily grew crops on an industrial scale. Reforms led to the establishment of production cooperatives for former hacienda workers, followed by a gradual redistribution of this land into small plots of less than one hectare. Initially, former hacienda workers and their families planted these plots to enhance food security by using crops such as maize and sweet potato. More recently, commercial vegetable production and livestock raising are increasingly common.

The neighbourhood of Carapongo rests at approximately 200 m above sea level in the lower zone of the Rímac River watershed. The community encompasses an area of 400 ha and has more than three times as much cultivated land than an average residential area. The Rímac River is the source of irrigation for Carapongo's agricultural plots. The water is conveyed through a system of irrigation channels which permit farmers to produce three or four harvests annually. The most commonly grown vegetables are beets, lettuce, cabbages, turnips, celery and radishes (Figure 3.3). Agriculture

Figure 3.3 *Urban agriculture in Carapongo*

constitutes an important strategy for obtaining income; however, it has developed in a context of unregulated urban growth, limited access to markets and restrictive policies.

Urban agriculture: a sustainable livelihood strategy

In order to analyse UA as a livelihood strategy, this section is divided into two parts. First, the types of assets of urban households involved in agriculture are discussed. Second, the different gender roles of men and women within the household are examined in order to present a more nuanced understanding of UA in Carapongo.

Assets of households practising urban agriculture

As noted above, urban producers use a variety of assets, which they combine in order to deal with risks and vulnerabilities. These assets are divided into five categories: natural, physical, human, financial and social.

Natural capital

Access to land – The process of urbanization is having a serious impact on Carapongo as agricultural fields make way for urban development. Of the 289 ha used for agriculture in 2002, 213 ha remained by 2006: a 26 per cent reduction. On the other hand, for the same period, the area devoted to housing increased from 113 to 197 ha, representing a 75 per cent increase (Figure 3.4).

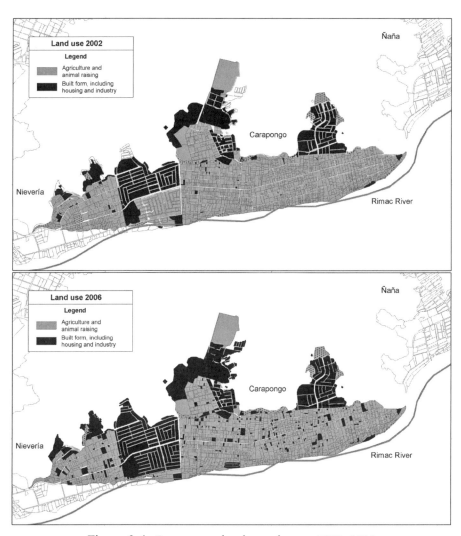

Figure 3.4 *Carapongo land use change: 2002–2006*

The study found that 37 per cent of UA producers in Carapongo occupy land that they own, which has an average area of 0.72 ha, whereas 19 per cent occupy rented land (average area 0.61 ha). Seven per cent of farmers occupy land located on the border of the river (average area 0.87 ha). This last category of farmers are farming illegally and are known as *pocesionarios*. Seventeen per cent of producers occupy land with a combination of ownership and rental, with an average area of 0.96 ha per holding.

Access to water – The principal water source for irrigation is the Rímac River, which runs through the city of Lima. The river nurtures an ecosystem in the urban and peri-urban areas that generates both benefits and risks for people. One of these risks is water pollution, which greatly affects agricultural producers of Carapongo, because the edges of irrigation channels have become deposits of solid and liquid wastes from nearby residents. Where quality requirements for food are high, such pollution prevents urban producers from accessing new markets. To access irrigation water, farmers pay a fee of around US$62.42 (206 nuevo sol/ha per year, where US$1 = 3.3 nuevo sol) to the water user association (Junta de Usuarios del Rio Rímac – JUR) that manages the resource.

Physical capital

Animal breeding – Animal husbandry represents physical capital for households practising UA. The purchase and raising of animals is a source of savings, as animals can be sold to obtain access to financial capital. In the area studied, 49 per cent of the farmers raise sheep, 42 per cent raise guinea pigs and 21 per cent raise pigs. Forty-two per cent of farmers raise chickens and 25 per cent reported that they raise ducks. Only sheep and pigs are sold for income (Figure 3.5).

Access to roads and transportation – Carapongo has easy access to transportation routes, which makes it possible for produce to reach the principal urban markets rapidly. Farmers use various strategies for transportation of their harvest. One way is to rent pick-up trucks or trucks. In such scenarios, two or three producers join together to share the transportation expenses. Larger producers use their own private transportation. Motorcycles are also commonly used by producers who sell their farm produce to an intermediary, who is then in charge of transporting the goods to market by truck (Figure 3.6).

Means of production – Urban producers mainly use the horse-drawn plough to work the land. Heavy machinery such as tractors is rarely used

Figure 3.5 *Livestock farmer in Carapongo*

Figure 3.6 *Different ways of transporting food from the field to urban market*

(Figure 3.7). On average, the majority of farmers have two backpack sprayers to fumigate the crops. The irrigation system is primarily comprised of bed and furrow irrigation and very few farmers utilize sprinkle irrigation. The urban production of vegetables is characterized by an intensive use of labour, mainly carried out by family members. It is common to find whole families

Figure 3.7 *Using traditional tools for soil management*

participating in agricultural tasks, from seeding to harvest (Figure 3.8). Contracting local labour is also common. Men are hired for certain jobs (washing produce, loading sacks and transporting), while women are employed for seeding, weeding and harvesting. The daily wage is approximately US$5 for men and US$4 for women.

Access to housing – More than half of the urban farmers interviewed stated that they owned their homes. In many cases, however, ownership does not have legal backing: 61 per cent of the people interviewed in this study do not have title to their property. According to de Soto (2000), 53 per cent of homes in the cities are 'extralegal' dwellings, that is, they are traded in an informal market without formal legal status. Further, de Soto points out that the lack of legality of these assets makes it difficult for households to generate financial capital because people are unable to access credit.

Human capital
Local knowledge and training – Urban farmers' knowledge is transferred in various ways: from generation to generation, from parents to children, and by the knowledge acquired over time through decisions made along the way thanks to the adoption of new and innovative ideas. In this zone, 75 per cent of farmers have come from rural areas (ENCUESTAS, 2004). They arrive with knowledge of how to cultivate land and raise animals. Others who do not

Figure 3.8 *Families participating in agricultural tasks*

have this traditional knowledge have had to learn it while living in the urban zone. Many families grow corn, potatoes and beans, and raise guinea pigs – a traditional activity in rural areas. Urban farmers also practise weed control, companion planting and use animal dung for fertilizer.

The formal educational level of farmers is low, with only 18 per cent of those surveyed having received any training. Nonetheless, on many occasions, farmers expressed a lack of interest in participating in training workshops offered to them. This is symbolic of an ongoing lack of trust directed towards any external agents who attempt to intervene.

Health status – Urban farmers in Carapongo have adequate access to health care because of good proximity to health centres and hospitals. Still, producers are exposed to health risks and in particular to gastrointestinal diseases, mainly because of polluted water. Some studies (TECNIDES, 2002) indicate that the water in this community is highly contaminated by fecal coliforms, parasites and other pathogens that surpass acceptable limits established by health legislations, both for drinking and irrigation water.

Perception of poverty – Low self-esteem, lack of self-confidence and the perception of poverty impede the ability to find solutions to one's problems and, by extension, the ability to develop one's 'human capital'. This study attempted to identify some issues facing farmers in Carapongo. To this end, one participatory workshop involved a discussion concerning whether farmers considered themselves to be poor or not. Results from the six participants are presented in Table 3.1 and an interesting trend was discovered: most farmers do not consider themselves to be poor. Factors influencing this personal

Table 3.1 *Perception of poverty*

Farmer	Sex	Rich	Medium	Poor	Very poor	Why?
1	M			X		I don't have land
2	F		X			I can access food, we have land
3	M		X			I have land, but I don't have enough
4	F		X			I can access land and food
5	F		X			I have other incomes
6	M		X			I have access to basic services

assessment include their access to land, crops and animals, all of which give food security to the family and strengthen self-esteem. On the other hand, feelings of being vulnerable and poor were expressed in the case where the participant did not have access to land.

Financial capital

Financial assets are the most significant limiting asset of poor people. Financial assets are also one of the most important as they can be used to leverage other types of capital. To speak about income and expenditures with the urban farmers was quite a delicate subject because they are not open about this issue; however, in the study we made an approximation of these financial components. This study identified that 51 per cent of the population in Carapongo only receive income derived from agriculture. For the analysis of the income variable, an approximation of the availability of cash to meet monthly expenditures is shown. In this study, 28 per cent declared that their families have less than US$150 available for their monthly expenditures; 41 per cent had more than US$150 but less than US$300; and 31 per cent had equal to or more than US$300 a month. Twenty-eight per cent of families with an average family size of five persons who live on US$150 a month – less than US$1 per person/day – indicates extreme poverty (Table 3.2).

Another important financial indicator is the proportion of total expenditures that are made on food. For 64 per cent of urban producer families in Carapongo, food expenditures comprise more than 50 per cent of their total expenditures. For the remaining 40 per cent of the families, food expenditures represent more than 70 per cent of their total household expenditures.

Table 3.2 *Income range and per capita income*

Income range	N	(%)	Per capita income	N	(%)
<US$150	34	28	<US$1 per day	32	26
<US$300	50	41	<US$2 per day	51	42
≥US$300	38	31	>US$2 per day	39	32
Total	122	100	Total	122	100

Loans and credits – Access for formal credit from banks is limited, and only 14 per cent of the producers reported having received bank loans. The remaining 86 per cent have not received any loan. Thirty-five per cent of all of the farmers surveyed have received credit (credit being different from a loan). For these farmers that have received credit, the main sources are stores that sell agricultural products and fertilizer, and other small enterprises. The principal reason for asking for financial assistance was to cover the input costs of growing vegetables. Lack of access to credit restricts the growth of their businesses and illustrates an important problem associated with smallholder farming in urban areas.

Savings – There does not seem to be a very broad range of options for saving money for the population living in Carapongo. Interviews revealed there is no custom of formal saving (in banks, for instance). Some people indicated that they save by keeping money safe at home or by purchasing animals that will be sold later.

Social capital

Carapongo is made up of a population who are originally from various regions and who grow different crops. For Durston (1999), common socio-cultural elements such as outlooks and values, memory of historical events, religion and myths, identity, rules of relationship, principles of reciprocity, as well as the social phenomena of neighbourhood and friendship, comprise the necessary precursors for the formation of social capital. He identified four levels of analysis: the individual; closed small groups, where there is shared experience and a high degree of trust; the wider community where there are common interests and objectives; and external relationships and interactions.

After noting Durston's constructs, four categories of social capital in Carapongo were identified: individuals; family or closed groups; community groups; and 'exogenous' groups, such as local government:

1 Individual actors, especially community leaders, have access to networks and contacts with other groups that permit them a greater control of information and resources.
2 Family and/or closed groups have kinship ties and linkages that influence their actions. This is especially true among those coming from rural areas who share land and help new arrivals to establish their housing and employment. Furthermore, in many cases it was discovered that activities in the field are performed by the members of the family.
3 Community groups are an important source of social capital and include organizations such as farmers' associations and groups that are initiated with the purpose of defending property rights both of land and housing. This research found that in some cases, the leadership of these groups was not completely trusted. The leadership of the Water Users Association was cited as more trustworthy than leadership of other groups.
4 Exogenous groups such as, for instance, local government are characterized by very weak ties and a high degree of mistrust within the community. An

example of conflict includes the issue of how lands are zoned in Carapongo. Farmer and community groups are advocating for the right to conduct agriculture, while local authorities do not permit such a classification.

Other indicators of the social capital are indexes of confidence and network formation.

Indexes of confidence – The strength of social ties can be quantified using a 'confidence index' – essentially, a way of quantifying different degrees of confidence between different actors. With regard to the index of confidence required to form an organization, the highest degree of trust is for neighbours (36 per cent), then family members (22 per cent). However, when enquiries were made about confidence in different institutions, the local government was rated poorly, with a 60 per cent level of mistrust, while a higher amount of confidence was shown in the Irrigation Board (with 35 per cent average confidence and 15 per cent high confidence).

Networks and other forms of organization – Urban farmers participate in organizations that have been formed primarily because of the presence of common interests in specific areas (such as water and land). To date in Carapongo no networks have been formed involving inter-sectoral or comprehensive approaches. Working with others in networks involves being part of a process where information is exchanged, new knowledge generated, experiences strengthened, resources exchanged, practices integrated and replicable models built for other projects. Lack of networking is a limitation since it impedes the ability to take advantage of opportunities generated by the market. Lack of networking also makes negotiating with other agents of the production and consumption chain more difficult.

Gender division of labour within households

We examined the gender division of labour to look at the roles of men and women in UA. Among other gender issues, this section examines gender division in access to and control over assets (with respect to control over the benefits of production) and productive resources. The survey of 125 farmers included questions related to gendered labour responsibilities, reproductive responsibilities and control over means of production.

Division of labour in agricultural work – One of the more interesting findings relating to the division of labour between men and women in UA is the frequency with which tasks are shared. Primary responsibility for agricultural tasks is summarized in Table 3.3. In only two types of tasks – land preparation and pest control – do men clearly possess sole responsibility. Women are generally responsible for these tasks only when they are single, widowed, or under other special circumstances. Where the task of land preparation is shared, women help to prepare vegetable beds that cannot be done by ploughing alone. Many other tasks are more frequently shared than defined specifically as male or female roles. Although men play a stronger role in vegetable production, this does not carry through to post-harvest and marketing activities, where women clearly play a bigger part. This is because

Table 3.3 *Primary responsibility for agricultural tasks of men and women in Carapongo (n = 125)*

Activities	Men (%)	Women (%)	Shared (%)	Neither (%)
Crops				
Land preparation	78	3	16	3
Planting	36	6	55	3
Fertilization, weeding, hilling up of soil, irrigation	30	6	62	2
Pest control	87	3	6	4
Harvesting	14	5	75	6
Livestock				
Raising small animals	6	23	28	43
Raising larger livestock	7	12	25	56
Purchase of inputs	46	22	30	2
Marketing products	23	41	36	0
Household labour	4	76	15	5
Day labourer	14	3	11	72
Off-farm activities	19	11	8	62

women are considered better and tougher negotiators. Women are also actively involved in caring for livestock, including feeding, health care and marketing. For some types of livestock such as poultry, women are principally responsible in almost two-thirds of all cases (Figure 3.9).

Domestic responsibilities – Women are totally responsible for fulfilling the role of family caretaker and are responsible for the raising of children and the maintenance of the household. Moreover, women must deal with the 'double day'. An activity profile developed with women in Carapongo illustrated that women have to combine a large number of activities during the day when they are at home, before and after going to the field. The work at home in the evening is similar to that done in the morning, but women consider it to be heavier because at the time they are very tired. The survey found that, on average, women spend seven hours working in agriculture activities and another eight hours in household activities, whereas men spend nine hours working only in agricultural activities. Women also undertake household management of waste. For instance, in the course of the discussion groups, many women identified the need to learn how to better manage and recycle organic wastes to produce nutrient-rich fertilizer, including the management of wastewater from household sewage and how it could be safely applied for use in vegetables production.

Control over the benefits of production – In group work, participants specified the degree of access and control they believe they have over different productive resources and the benefits of UA (Table 3.4). The discussion group itself determined the definitions of 'control' and 'access'.

Figure 3.9 *Women play a central role in urban agriculture in Carapongo*

There were many parallels between men and women in reporting on their tenure security (Table 3.5).

None of the types of tenancy mentioned above are gender-specific. Nevertheless, men are more likely to assume responsibility for crop production in Carapongo. Of the 125 households surveyed, men were

Table 3.4 *Access to and control over resources in Carapongo*

Productive resources	Access*		Control*	
	Men	*Women*	*Men*	*Women*
Land	●●●	●●	●●●	●
Water	●●●	●●	●●●	●●
Inputs	●●●	●●●	●●●	●●●
Social capital (credit, participation)	●●●	●●	●●●	●●
Human capital (training, information)	●●	●	●●	●
Benefits of production (i.e. income)				
Income from sale of vegetable production	●●●	●	●●●	●
Income from sale of animal production	●●	●●●	●	●●●
Income from labour (from off-farm activities)	●	●●●	●	●●●

*●●●, complete access/control; ●●, partial access/control; ●, limited or no access/control.

Table 3.5 *Land tenure classified by gender and reported by the household head*

Tenancy	Men (n = 88)		Women (n = 37)	
	Area (ha)	Percentage	Area (ha)	Percentage
Ownership	0.72 ± 0.42	38	0.72 ± 0.67	36
Rental	0.68 ± 0.39	16	0.52 ± 0.40	28
*Posecionarios**	0.82 ± 0.98	9	0.84 ± 0.67	6
Ownership + rental	1.0 ± 0.46	17	0.34 ± 0.22	5
Ownership + posecionarios	1.52 ± 0.80	4	1.54 ± 1.37	11
Other types	0.85 ± 0.10	16	1.06 ± 0.62	14
Total		100		100

Posecionarios are those farmers farming land illegally (i.e. on land that has a different designated use).
Source: Survey data.

identified as being primarily responsible for the farm in 70 per cent of cases, and women in the remaining 30 per cent. Of the men responsible for the farm, 38 per cent lived on their own land, but less than half had a formal title. Among the 30 per cent of households where women were mainly responsible for farming, the pattern was the same: although 36 per cent owned their own land, only 49 per cent held formal title to it. The principal problems identified by farmers include contaminated water and the resulting contamination of food, particularly vegetables.

Exactly who has access to and control over inputs for crop and livestock production was related directly to the purpose of production: either for market or for home consumption. Both men and women invest significant inputs for commercial production (i.e. cash crops and animals for sale), whereas it is mostly women who perform subsistence production (small-scale plantings of root and tuber crops, beans, green maize and herbs, along with small animals such as poultry and guinea pigs). The purchase and use of pesticides is mostly done by men, partly because of the physical exertion involved, but also because of the risk of contaminating children and food if women handle pesticides. In some cases, hired labour is used to replace family labour in the use of pesticides and other inputs. Male farmers have better access to agricultural credit because of more frequent interactions with suppliers. In almost 90 per cent of cases, men are responsible for crop protection: both purchasing and applying chemicals.

Most community organization activities among the farming population in Carapongo are male dominated. Men predominate in existing associations of water users. Women make up only 10 per cent of membership in the committees, but play an important role at the household level with domestic water management in Carapongo.

In terms of gender differences and human capital, both men and women lack access to training and information about crops and livestock husbandry practices, though women are at a much greater disadvantage. Only 22 per cent of the farming population had received agricultural training, and of these, 86 per cent were men and 14 per cent were women. Women were particularly interested in learning more about basic methods for treating animal health problems, whereas men were more interested in information about sources of credit and government training programmes.

Decision making – Women predominate in the marketing of farm products, while both men and women share in the decision to sell and in the control of the proceeds of sales. Where a woman has control over land, she most often has the right to decide on the sale of produce. She also maintains control over the proceeds and, as in one case recounted by a participant, a woman might choose to use the proceeds from the sale of inherited land to pay rent on other land. On the other hand, men are more commonly in control of property and they generally have more authority in decisions about selling land.

Precisely how decisions are made within the family depends on how farming tasks are divided. One of the most important dimensions in the organization of farming systems is the division of labour based on gender. Both the position of individuals within the household and the division of labour affect an individual's knowledge of the crop–livestock system. This, in turn, determines who makes the decisions. Commercial farmers make more decisions alone as well as fewer in consultation with family members; however, farmers producing for home consumption or who operate non-land-owning households with animals make fewer decisions alone and more in consultation with family members. The extent of consultation also depends upon how farmers perceive the importance of the decision. A summary of decision-making is shown in Table 3.6.

Outcomes of urban agriculture in Carapongo

Even though farmers who were interviewed during the study generally declared that UA is not profitable, the majority obtain some income from the sale of crops or animals which makes it possible for them to make a subsistence living. Those farmers who have secure access to the land (either owned or rented) generally have more opportunities to earn a higher income and attain a better quality of life. At the same time, some farmers generate employment for other people, mainly those who live in the vicinity of Carapongo or along the river bank (marginal zone). Many farmers also reported that, thanks to the income generated by UA, they have managed to get more land and build their house with more robust construction materials. Furthermore, they have managed to obtain education for their children, some of whom reach a technical and/or university education.

Most agricultural produce – lettuce, turnip and beets – is sold by farmers and not eaten in their homes. However, in the orchards space is reserved for seeding other products they usually eat. Furthermore, in small-animal husbandry meat and eggs represent other food sources.

Table 3.6 Decision-making matrix in Carapongo households

Decision	Decision process			Comments
	Men dominate	Equal influence	Women dominate	
Inputs				
Who decides how the family labour will be used?	•			When men work off-farm, women spend more time in the field or hire labour for the farm
Who decides what inputs to buy?		•		The crop, animals and type of use (for food or for cash) determine who buys and what to buy
Production				
Who decides which crop to grow?	•	•		Depends on demand for the crop and on capital. Women landowners/renters might decide on which crops to grow
Who decides when to harvest?	•	•		When the crop is ready for harvest there is always some flexibility
Who decides on whether the family should sell for cash or consume as food production (crop/livestock)?				Depends upon the land, water, capital available, labour and knowledge of the crop/animal production
Who decides on the number of animals to buy?			•	Depends upon the household labour and capital available. Example, a case of a man deciding to buy ducks
Marketing				
Who decides what part of the harvest is sold and how?		•	•	Women decide on how much of the production (vegetables) is required for market and for household consumption, and how much can be taken to market. Both men and women choose to harvest a few heads of lettuce for eating
Who decides when animals will be sold?		•	•	Women usually make this decision but men can influence it, specifically for commercial production such as pigs
Who decides what animal products will be sold and how?		•	•	The only animal product is pig meat
Investments				
Who decides to buy or rent additional land?	•	•		Men and women make decision, but men dominate in this area
Who decides to buy more animals?		•	•	Depends on capital available and space to keep animals. Women dominate this decision
Reproduction				
Who decides whether a child goes to school or not?		•		Both make this decision
Who decides on going to a doctor?			•	Women dominate the decision

Source: Results from workshop discussions.

There is also evidence that the practice of UA strengthens family ties which, in some cases, make access to housing or land possible. Also, many farmers, particularly those of rural origin, are supportive of others who share their rural roots. This has enhanced a sense of community amongst farmers in Carapongo. Moreover, farmers feel that despite various inherent problems associated with an agricultural lifestyle, they fulfil an important role in feeding the city. They tend not to feel poor, because they have access to food, work and education for their children.

It was clear that urban farmers recognize that their lifestyle is risky because of exposure to water pollution, solid waste and agricultural chemicals. Another problem they emphasized relates to their exposure to market forces that are unpredictable. Farmers identified a number of reasons: price instability, the lack of information networks, the lack of farmer organization and fear that the quality of their products may be substandard. Finally, many farmers are increasingly concerned that agriculture will no longer be a profitable activity due to the low price of crops as compared to the high cost of inputs.

The assets and strategies developed by urban agricultural households in Carapongo are influenced by many complex institutional and political relations, both of which positively and negatively affect farming families.

In metropolitan Lima, the main wholesale market does not possess an adequate infrastructure and is currently saturated with an excess of agricultural products. This has emerged mainly due to a lack of planning and ineffective regulations, both of which contribute to instability in crop pricing. This instability has negatively affected farmers' household economies because they do not have other markets where they can sell their products. As noted above, the local government does not recognize UA as an important activity when developing city planning. Therefore, despite the existence of a local 'municipality law' that promotes green space within the city, local decision-makers give priority to urban development in favour of UA. This situation generates conflict between the urban farmers and municipal authorities.

Perceptions and prospects of the main stakeholders

Farmers perceive UA as an activity that concerns neither the local authorities nor the central government. Most feel betrayed by failed promises pledged by politicians and believe that the prospects for improvements are uncertain. As a result, most farmers report being pessimistic about their future. They fear the time when they no longer have land to work because of increasing urbanization. Despite this unfortunate situation, older farmers cannot imagine themselves performing any other type of economic activity because agriculture is all they know and at their age they believe they will be unable to find another job. Hence, they expect to continue working in agriculture, possibly in another location. On the other hand, younger farmers are not as concerned and they are willing both to take risks and to undertake another business if necessary. The following quotes represent farmers' various feelings:

In a couple of years all this land will disappear, because all this is being populated. Everything was land but now it's all houses. Samuel Yupanqui, Carapongo farmer.

In five years this surely is going to be urbanized because people will no longer want to plant vegetables. Edgard Palacios, Carapongo farmer.

The land is no longer agricultural, this is already housing. I already have noticed that, because the council has forced us to accept that this is now an area of urbanization. Pasión Reynoso, Carapongo farmer.

Five years from now, what can I do – work on the farm until I die? As I say, when you are already old, no one can give you work. Edmundo Vicuña, Carapongo farmer.

Interviews with responsible authorities suggest a dramatic difference in perspective. Their perception of UA is that it is a marginal sector. They give no importance to the conservation of agricultural areas in the district. Instead, their main priority is the physical development of urban infrastructure. From their point of view, the concept of environmental care refers to issues related to the reforestation of the Rímac River and the maintenance of the city's parks and gardens. In general, they do not fully understand the concept of UA and its importance within the city. The authorities believe that in the medium term agricultural activity will disappear in this area and everything will be urbanized.

There are several key elements that enhance the disagreement between these two main parties. For farmers, there is a high level of mistrust directed at the municipal management as well as a rejection of their authority. Three main conflicts exist:

1 A subject that causes particular conflict is public health, specifically because the farmers do not agree to having to pay municipal taxes for services that are not provided. For example, although local authorities collect taxes for garbage collection in agricultural areas, the municipal garbage truck does not always come.

2 Another point of disagreement with the municipality is sanitation. The farmers say that the municipality supports urbanization without considering that the area is mainly agricultural. Because they are not yet urbanized, these agricultural zones do not qualify for the necessary sanitation infrastructure, so there are no sewerage systems in the agricultural lands. As a result, settlers use the irrigation channels as sewers and garbage dumps. This behaviour contaminates the water use for irrigation purposes, and can damage the farmers' health.

3 Conflict also exists relating to classification of land use in Carapongo. The Central Municipality of Lima has designated the area as being an urban

residential zone, a classification which naturally conflicts with the perspective of the farmers.

CONCLUSION

This research project is one of the first initiatives in the country linking UA with the livelihoods framework. The analysis presented in this chapter provides a holistic and integrated view of the processes by which people achieve (or fail to achieve) sustainable livelihoods. This study has identified the principal problems and opportunities for UA which is being used in a larger initiative, Agricultores en la Ciudad (Farmers in the City) of the Urban Harvest Program of CIP. The initiative is designed to formulate strategies for intervention in the UA sector in Carapongo.

Furthermore, this research contributed to the development of policy guidelines developed by Carapongo's local government. These guidelines are manifested in a municipal document, *Regulation of Organization and Functions*, which the management regulates. The document outlines the functions for the new Department of Urban Agriculture, a department that opened within the municipality with management support from the Urban Harvest Program of CIP.

ACKNOWLEDGEMENTS

I would like to gratefully acknowledge the financial support from AGROPOLIS. I thank my thesis adviser Professor Joel Jurado of Pontificia Universidad Catolica de Lima of Peru and my supervisor Dr Gordon Prain, coordinator of the Urban Harvest Program. Many thanks to Lina Salazar for her comments on the final report. And finally, a special thanks to all of the urban producers and local authorities who collaborated during the fieldwork.

NOTE

1   Luis Maldonado Villavicencio, MSc. Social Management, Pontifical Catholic University of Peru. Address: Calle 51 no 280 Urb. Corpac San Isidro, Lima, Peru; tel: +51 12260309; email: l.maldonado@cgiar.org

REFERENCES

Antweiler, C. (2000) 'Urban knowledge for a citizen science, experience with data collection in eastern Indonesia', presented at the Association of Social Anthropologists (ASA) Conference 2–5 April 2000, Participating in Development, London, UK

Armar-Klemesu, M. (2000) 'Urban agriculture and food security, nutrition, and health', in *Growing Cities, Growing Food: Urban Agriculture on the Policy Agenda*, German Foundation for International Development (DSE), Feldafing, Germany

Chambers, R. and Conway, R. (1992) 'Sustainable rural livelihoods: Practical concepts for the 21st century', Institute of Development Studies (IDS), Brighton, UK, Discussion Paper No 296

Durston, J. (1999) 'Construyendo capital social comunitario', Revista de la CEPAL 69 (LC/G.2067-P), Santiago, Chile, December

ENCUESTAS (2004) Encuestas realizadas en Carapongo. Linea de Base del Proyecto Agricultores en la Ciudad. Cosecha Urbana-Centro Internacional de la Papa (CIP), Lima, Peru

Hovorka, A. J. (1998) 'Metodología para el análisis de género para la investigación sobre agricultura urbana', *Cities Feeding People Series* 26

Instituto Nacional de Estadística e Informatica (INEI) (2002) Encuesta Nacional de Hogares (ENAHO) IV, Trimestre 2001, Lima, Peru

Martin, A., Oudwater, N. and Meadows, K. (2000) 'Urban agriculture and the livelihoods of the poor in Southern Africa: Case Studies from Cape Town and Pretoria, South Africa and Harare, Zimbabwe', Paper presented at the International Symposium 'Urban agriculture and horticulture: the linkage with urban planning', 7–9 July 2000, Berlin, Germany. Sponsored by TRIALOG, Vienna, Austria; Humboldt-University, Berlin, Germany; HABITAT, Havana, Cuba

Maxwell, D. (1994) 'Internal struggles over resources, external struggles for survival: urban women and subsistence household production', paper presented at the Association for African Studies, 3–6 November 1994, Toronto, Canada

Mbiba, B. (1993) 'Urban agriculture, the poor and planners, a Harare case study', 10th Inter-schools Conference on Development, 29–30 March 1993, Philips House, The Bartlett Graduate School, University College London, London, UK, pp129–135

Mianda, G. (1996) 'Women and garden produce of Kinshasa: The difficult quest for autonomy', in Ghorayshi, P. and Belanger, C. (eds) *Women, Work, and Gender Relations in Developing Countries*, Greenwood Press, Westport CT, pp91–101

Mougeot, L. (2000) 'Urban agriculture: definition, presence, potentials and risks', in Bakker, N., Dubelling, M., Gründel, S., Sabel-Koschella, U. and de Zeeuw, H. (eds) *Growing Cities, Growing Food. Urban Agriculture on the Policy Agenda*, German Foundation for International Development (DSE), Feldafing, Germany

Mudimu, G. D. (1996) 'Urban agriculture activities and women's strategies in sustaining family livelihoods in Harare, Zimbabwe', *Singapore Journal of Tropical Geography*, 17(2), pp179–194

Nugent, R. (2000) 'The impact of urban agriculture on the household and local economies', in Bakker, N., Dubelling, M., Gründel, S., Sabel-Koschella, U. and de Zeeuw, H. (eds) *Growing Cities, Growing Food: Urban Agriculture on the Policy Agenda*, DSE, Feldafing, Germany

O'Reilly, C. and Gordon, A. (1995) 'Survival strategies of poor women in urban Africa. The case of Zambia'. NRI Socio-economic Series 10, Natural Resources Institute, Chatham, UK

Rakodi, C. (1988) 'Urban Agriculture: Research Questions and Zambian Evidence', *The Journal of Modern African Studies*, vol 26, no 3, pp495–515

Rakodi, C. (1991) 'Women's work or household strategies?', *Environment and Development*, vol 3, no 2, pp39–45

Sanderson, D. (2000) *Cities, Disasters and Livelihoods*, CARE, United Kingdom. Also available at www.livelihoods.org/cgi-bin/dbtcgi.exe (accessed 20 November 2007)

Singh, N. and Gilman, J. (1999) 'Making livelihoods more sustainable', *International Social Science Journal*, vol 51, no 4, pp539–545

Soto de, H. (2000) *El Misterio del Capital*, Empresa Editora El Comercio S.A., Lima, Peru

TECNIDES (2002) 'Uso de tecnologías no convencionales para el abastecimiento de agua potable y letrinización en zonas rurales y/o urbano marginales', Programa APGEP-SENREM, Convenio USAID-CONAM, Lima, Peru

UNDP (1996) 'Urban agriculture: food, jobs and sustainable cities', publication series for HABITAT II, vol 1, UNDP, New York

4

The Social and Economic Implications of Urban Agriculture on Food Security in Harare, Zimbabwe

Charity Mutonodzo[1]

INTRODUCTION

For decades, poverty, food insecurity and malnutrition in Africa were viewed as largely rural problems (Maxwell et al, 2000); however, the population of many African countries, including Zimbabwe, is becoming more urban. The urban population in Africa grew from 27 to 38 per cent between 1980 and 2000, and is expected to reach nearly 50 per cent by 2020 (WRI, 1999). Unfortunately, the urbanization of the continent does not mean economic opportunity and prosperity for the majority of Africans. On the contrary, global poverty is becoming more African, more urban and more feminine. Fifty per cent of the world's poor and 40 per cent of Africa's poor live in urban areas (Rabinovitch, 1999).

Earlier research in Zaire has shown that 60–80 per cent of the total household budget of the poor is spent on food (Tabatabai, 1993). This finding makes it likely that urban poverty will be manifested at least in part as a problem related to food security. As food security continues to worsen in urban areas, households probably will turn to urban agriculture (UA) as a means of coping. Nevertheless, research on this topic is still relatively new despite an increasing but still limited recognition of UA (Mougeot, 2006). Part of the reason may be that many local governments and municipal decision-makers continue to view agriculture as essentially 'non-urban', which blinds them to UA's contributions to food security and socio-economic improvements. Hence, there is an urgent need to apply food-security monitoring approaches developed in rural contexts (FEWSNET, 2001) to the urban environment.

OBJECTIVES AND HYPOTHESES

Definition of terms

These are the main terms used in the paper requiring definition:

- Urban agriculture (UA): Agricultural activities (farming) in an urban setting. As defined by Mbiba (1995), it is 'the production of crops and/or livestock on land which is administratively and legally zoned for urban uses'. Mougeot (1995) defines it as 'the production, processing, marketing and distribution of crops and animals and products from these in an urban environment using resources available in that urban area for the benefit largely of the residents of that area'.
- Food security: The United States Agency for International Development (USAID, 1996) defines food security as a situation where all people at all times have both physical and economic access to sufficient food to meet their dietary needs for a productive and healthy life. However, for the purpose of this limited study, the researcher opted to use a narrower measurement of household food security. The indicator of food security was based on energy intake, and people who met 70 per cent of their energy requirements from cereals were considered food secure. Energy adequacy in the current study is taken as sufficiency of kilocalories (kcal) consumed in relation to requirements.
- Vulnerability: Moser (1996) defines vulnerability as the 'insecurity of the well-being of individuals, communities or households in the face of a changing environment'. Vulnerability can include the risk of, or susceptibility to, food insecurity and can result from either chronic or transitory conditions.

Objectives

The research had five objectives, which were designed to lead to an understanding of the socio-economic implications of UA in food and livelihoods security:

1   To characterize sampled urban households in terms of their vulnerability, poverty levels and gender of household's head among those that engage in UA compared with those that do not. This objective remained unchanged and was fulfilled during research.
2   To assess food security and livelihoods in terms of UA participation and other demographic, socio-economic and geographical characteristics.
3   To measure production performance of urban farmers compared with smallholder farmers throughout Zimbabwe.
4   To identify key factors accounting for variance in crop production performance and crop choices.
5   To explore policies and institutional innovations that are likely to safeguard and enhance the contribution of UA to the food security of the vulnerable urban population.

Unfortunately, a number of the objectives had to be modified during the research process due to a number of external influences. For instance, objective 3 had to be dropped due to changes that were brought about by Operation Restore Order, a project started on 19 May 2005 by the Government of Zimbabwe to clean up its cities. The operation resulted in the destruction of homes, business premises and vending sites, including most of those who were practising UA who were interviewed in April and early May 2005. Those affected moved to stay with friends and family or found temporary shelter in churches. Because many of the displaced persons in this sample were scattered in different locations and were highly mobile, it became impossible to measure their production performance.

Meanwhile, the fourth objective was also modified to mirror the changes brought about by Operation Restore Order because it was impossible to assess crop performance and farmer choices. This objective, therefore, became more generic, being restated as: To identify key factors accounting for variation in food security and livelihood status in an urban setting.

METHODS

A livelihoods-based vulnerability analysis framework, based on the assessment used during the Zimbabwe Vulnerability Assessment (ZIMVAC, 2003), was applied to guide data collection and analysis. This framework seeks to clarify the mechanisms by which people obtain access to food and other essential resources and services within communities, by offering a framework that allows for interpretation and monitoring of data such as market prices. It is a framework that provides a geographic and socio-economic basis on which to understand the strategies that people use to meet their basic needs (FEWSNet, 2005). Data sources included a secondary data review, household surveys, focus-group discussions and key informant interviews. The widely accepted minimum daily energy requirement of 2100 kcal per day for an average adult (FAO, 1993; WFP, 2000; UNHCR and WFP, 1997) was used as a benchmark for the study. However, since other nutrient requirements are such that 10–12 per cent of energy in the diet should be in the form of protein and at least 17 per cent of the energy in the diet should be in the form of fats, by difference 70 per cent of the daily energy requirements of 2100 kcal come from cereals. Food security is achieved when at least 70 per cent of the energy requirement is met.

Four focus groups for soliciting the community's perspectives on the division of labour, income and food sources, coping strategies and challenges faced were assembled. The groups consisted of 8–10 men and women. Each group included a mix of people who were and were not involved in UA. A gender-specialist facilitator guided the discussion to ensure it stayed on topic, while the researcher and a note-taker recorded the discussion. A number of participatory approaches were used: seasonal calendars for understanding seasonal aspects, timelines to identify events affecting well-being, proportional piling to determine relative magnitudes and Venn diagrams to depict the various groups' relationships with other institutions. The research was done with the participation of a number of teams within the departments of Agricultural Research and Extension and the Ministry of Agriculture (both the National Early Warning Unit and the Harare Urban Extension teams). Both work in the areas of UA, food security and livelihoods research.

A review of the available literature on UA in general and in Harare in particular was carried out, using both published and unpublished documents. Six weeks of observation and unstructured interviews with city authorities and households involved in UA were a starting point for the research.

Study sites

Harare Metropolitan Province is divided into four districts: Harare Rural, Chitungwiza, Epworth and Harare Urban. Harare Urban had the largest proportion (76 per cent) of the population in the 2002 national census. The present study took place in the Harare Urban district, which is made up of 48 wards spread over 27 suburbs. In this district, on-plot agriculture exists and was provided for in low-density suburbs such as Borrowdale. Meanwhile, off-plot agriculture is found across the city in suburbs of varying densities, with varying degrees of intensity and tenure regimes. Peri-urban agriculture (UPA) exists in areas surrounding the city and some of the lands are smallholdings zoned for agricultural purposes.

A total of 16 suburbs where most of UA was assumed to be taking place were selected for the study. Generally speaking, high-income residents tend to inhabit low-density areas, medium-income residents live in suburbs of all density types and low-income residents mostly inhabit high-density suburbs. The distribution of the 372 households among the selected suburbs is shown in Table 4.1.

A household survey which included a consent form was used to collect information from sampled households. It collected information on demographics, assets, support to UA, UA-related activities and formal and informal employment. It also collected information on coping strategies related to income, expenditures and consumption.

Survey sampling and sample size

The study used a two-stage sampling strategy, where the first stage was the ward or suburb. A list of these was obtained from the Central Statistics Office, representing wards used for the 2002 population census. These were deliberately chosen on the basis of residents' participation in UA. The second stage for sampling was households, chosen randomly by selecting people who were in their fields during the time of the interview.

The minimum sample size required was calculated using the formula (Equations 4.1, 4.2) for large samples given by Poate and Daplyn (1993):

$$n = \frac{z^2 c^2}{x^2} \qquad\qquad (4.1)$$

Table 4.1 *Distribution of sampled households by suburb name and type*

Suburb name	Suburb type	Sample size (number practising UA)
Greendale	Low density	19 (19)
Hatfield	Low density	22 (20)
Mabelreign	Low density	16 (16)
Mt Pleasant	Low density	19 (18)
Waterfalls	Low density	20 (20)
Tynwald	Medium density	20 (20)
Budiriro	High density	25 (20)
Dzivarasekwa	High density	26 (22)
Glen Norah	High density	25 (20)
Glen View	High density	26 (21)
Hatcliffe	High density	22 (22)
Kambuzuma	High density	25 (23)
Kuwadzana	High density	23 (16)
Mabvuku	High density	26 (25)
Tafara	High density	24 (23)
Warren Park	High density	30 (26)

where n is the minimum sample size required; z is 1.96, the value of z at the 95 per cent confidence interval; c is the variation within the population, which has been assumed to be 60 per cent; and x is the expected level of accuracy, which has been estimated at 6.1 per cent.

Therefore:

$$n = \frac{(1.90)^2(60)^2}{(6.1)^2} = 372 \tag{4.2}$$

Conducting the study

The questionnaire was pre-tested in three different areas of the city, modified and delivered by eight surveyors selected and trained by the researcher. The training sessions served to explain why the survey was being conducted, its relevance to urban and national development and the rationale for sampling. Surveyor training and pre-testing of questionnaires took place during a two-week period of November 2005. After training, enumerators received notes explaining the meaning, relevant concepts and definitions for each question, as well as the techniques for data collection. Enumerators helped translate the questionnaire into Shona, the language spoken in the study area. The survey was carried out over a period of two weeks in December 2006. Questionnaires were field-checked by the academic supervisor of this work and subsequently by the researcher.

Data management

Data entry took four weeks using the Statistical Package for Social Sciences (SPSS, version 10) for data entry. To minimize errors, a double-entry system was used and all discrepancies were corrected by referring back to the questionnaire. Data cleaning on the original data files was also conducted in SPSS (version 10). Tabulation and further analysis were done in SPSS versions 10 and 11. A total of 368 questionnaires remained from the original 372 after data cleaning.

Analytical techniques

Livelihood and food security analyses were conducted by looking at food and income sources, as well as amounts, expenditure patterns, coping strategies and required daily allowances compared with energy adequacy.

Statistical analyses

Statistical analysis used in this research (see below) involved descriptive statistics such as mean, median, proportions and odds ratios. These were used to compare those households practising UA with those that were not. Independent t-tests were applied similarly. Associations between categorical variables were investigated using chi-square tests. A P value of 0.05 or less was considered significant. A linear regression model was fitted to determine factors affecting food security. A binary logistic regression model was fitted for participation in UA, which has a binary outcome (yes or no). Odds were calculated for the categorical variables such as sex, educational level of household head and suburb type.

Food security analysis
Annual household food requirements

The widely accepted minimum required intake of 2100 kcal per day for an average adult was used as the benchmark during this research (WFP/UNHCR, 1997; WHO, 2000; WFP, 2000). This is the minimum requirement for adults against which people's access to food was compared. However, it is acknowledged that differences in requirements between households with the same total size, but different demographic composition, are masked by using this method. Based on the information from these studies, we assumed that cereals supplied 70 per cent of the required food energy, or 1470 kcal/person per day, for an average household member. This is equivalent to 148 kg of cereals per year. These energy requirements were disaggregated by age and gender, as is indicated in Table 4.2.

The total number of household members was calculated in the age and gender categories listed above. The number of people in each category was then multiplied by the quantity in kilograms of cereal required by that age and gender group in order for them to meet 70 per cent of their minimum cereal needs.

Table 4.2 *Household energy requirements by age and gender*

Age	Energy requirements (kcal/person/day)		Cereal requirements (kg/year) at 70% of energy	
	Male	Female	Male	Female
0–4	1320	1250	92.9	88.0
5–14	2175	1885	153.1	132.7
15–19	2700	2120	190.0	149.2
20–59	2460	1990	173.1	140.1
60+	2010	1780	141.5	125.3

Source: WFP/UNHCR, 1997.

Total cereals accessed in 2005–2006

Total cereals accessed by households in the 2005/06 marketing year was calculated by summing all sources of cereals:

- *Personal production consumed:* The amount of maize accessed – the most common cereal – was obtained by subtracting the amount sold, exchanged or given away from the total amount harvested from urban areas, plus the rural production consumed in urban areas. Tubers were converted to an equivalent amount of maize, then added to other cereals to give the total kilograms of cereal consumed from people's own production. The conversion factor used was 1140 kcal/3630 kcal = 0.31; that is, kilograms of tubers were multiplied by 0.31 to get the maize equivalent. This was divided by 12 to get the monthly personal production contribution.
- *Purchased food:* Information was provided by households related to the quantity of food purchased. This was converted into kilocalories for the study month.
- *Direct sources of food:* Cereals – maize, sorghum and millet, wheat, rice and maize meal – obtained from relatives, churches or other donors and other friends other than derived from personal production were calculated for eight months and monthly quantities obtained.

Food security status

Food security status was calculated using the contribution of all cereal foods available to the family's household calorie intake during the month of the survey. The quantity for each household per month as well as for each food item was calculated for the survey month. Then, using calorie requirements for different age groups, the ideal monthly energy intake for each household was calculated.

Limitations of the study

The study does not claim to be exhaustive and flawless, because there are some assumptions and drawbacks that one needs to be aware of when using the results (as shown below in Research Findings). Interpretation of results should be limited by the fact that these data are cross-sectional because of the single round of data collection. Also, consumption data do not capture seasonality or whether households experienced chronic or transitory food insecurity.

The household survey only recorded quantities of cereals and tubers accessed. It did not capture other foods consumed, such as legumes, meat, oils and fats or greens. It was assumed that 70 per cent of required energy is derived from cereals (1470 kcal). This presented two problems: first, how to determine the cut-off point; and second, whether it made sense, given variation across households as well as individuals' dietary requirements.

Another limitation is related to the survey sample. The sample was chosen purposively and is representative of those suburbs where UA is undertaken. As a result, the descriptive statistics reported in this study are not necessarily representative of all the households in Harare, which include neighbourhoods where UA is limited. Another important point is that a precursor of inclusion was that those people studied had to be occupying a household. This meant the homeless, as well as street children, were left out regardless of whether or not they were practising UA.

RESEARCH FINDINGS

Objective 1: Characterization of sample

Table 4.3 shows basic demographic characteristics of the sampled households (N = 368) by gender of head of household and participation in UA. A total of 69 households, or 18.8 per cent of the total sample, were headed by females.

Table 4.3 *Descriptive characteristics of surveyed Harare households by the gender of the household head and by participation in UA*

Characteristic	All households (n = 368)	Male-headed households (n = 299)	Female-headed households (n = 69)	Practising UA (n = 331)	Not practising UA (n = 37)
Mean age of head in years	46.9 (13.4)	45.9 (13.8)	50.9 (10.7)	47.7 (13.3)	39.5 (12.3)
Mean household size (persons)	5 (2.1)	5.2* (2.1)	4.6* (2.2)	5.1 (2.2)	4.2 (1.4)
Dependency ratio	0.94 (0.84)	0.91* (0.84)	1.04* (0.92)	0.94 (0.85)	0.90 (0.71)

Note: Numbers in parentheses are standard deviations.
*Significant at 5% level.

Table 4.4 *Factors associated with practising UA (logistic regression model)*

	Beta	Standard error	P value	Exponential beta
Suburb type: high density = 1, low density = 0	3.03	1.05	**0.004**	20.71
Gender of head of household: male = 1, female = 0	−1.50	1.38	0.28	0.22
Age of head of household (continuous)	−0.02	0.02	0.38	0.98
Educational level of head: 1 = primary, 0 = more than primary	2.07	1.03	**0.04**	7.92
House ownership: 1 = owner; 0 = non-owner	1.51	0.51	**0.003**	0.22
Household size (continuous)	0.28	0.14	**0.05**	1.33
Formally employed members: 1 = any formally employed, 0 = none	0.88	0.88	0.32	2.42
Informally employed members: any informally employed = 1, 0 = none	0.49	0.91	0.59	1.64
Constant	18.88	49.64	0.70	15.91

Overall percentage correct 89.9%. Significant results are shown in bold.

In this study, households headed by males were, on average, significantly larger than households headed by females.

Though female-headed households were smaller in size than male-headed households, they had significantly higher dependency ratios. Maxwell et al (2000) noted a similar finding in Kampala, Uganda, in the early 1990s. Households headed by females contained a much larger proportion of relatives and grandchildren than those headed by males. Female heads tend to be older (mean age 50.9 years) than their male counterparts (mean age 45.9 years).

Results for the logistical regression on UA are shown in Table 4.4. A correct prediction value of 89.9 per cent was obtained. This meant that the variables used were very good at statistically predicting the observed outcome of practising UA. Suburb type ($P = 0.004$), educational level of head ($P = 0.04$), household size ($P = 0.05$) and house ownership ($P = 0.003$) were significantly associated with participation in UA (see also Table 4.4).

Suburb type

Households in the high-density suburbs were more likely to practise UA than their counterparts from low-density suburbs, a finding consistent with ZIMVAC (2003), Mwakiwa (2004) and Mudimu et al (2005). From the odds ratio given by $\exp \beta$, those from high-density suburbs were 21 times more likely to practise UA than their low-density counterparts.

House ownership

Owner-households were more likely to practise UA than those who were staying in rented accommodation, which corroborates what Mudimu et al (2005) found. The odds ratio for this variable is 0.22, meaning that owner-households were approximately 22 per cent more likely to participate in UA compared with those in rented accommodations.

Household size

Larger households were more likely to participate in UA than small households. The Zimbabwe Vulnerability Assessment Committee makes the same observation (ZIMVAC, 2003).

Educational level of household head

Households where the head had up to primary level education were approximately eight times more likely to practise UA than those whose heads had educational levels beyond primary level.

Objective 2: Assessment of food security

Overall, based on the benchmark of having 70 per cent of the recommended 2100 kcal/aeu per day (where 'aeu' is adult equivalence unit) for Zimbabwe, 92 households, or 25 per cent of the sample, were considered to be food secure. The percentage of household energy requirements for food-insecure households increased with expenditure, from 66.9 per cent for the lowest expenditure quartile to 80.9 per cent for the highest. A rather surprising result is that the mean adequacy for female-headed households was higher than for their male counterparts. This was the case both for households that were food secure (107.3 per cent of caloric requirements met compared with 106.9 per cent) and for those that were food insecure (79.1 per cent compared with 73.8 per cent).

A summary of the relationships between food security and the practice of UA is given in Table 4.5. Among those that practised UA, about twice as many households (26.3 per cent) were food secure compared with 13.5 per cent among those not practising UA. Conversely, households defined as food secure and practising UA met 76 per cent of their requirements as compared to food-insecure households not practising UA, who met 65 per cent of their energy requirements.

Households whose household head contributed to UA on a full-time basis tended to meet more of their requirements (79 per cent of caloric requirements vs 66 per cent for households whose heads do not contribute to labour). Households represented with owners met more (79 per cent) of their requirements, compared with 70 per cent where housing is provided by employers (signifying less security of land tenure). Families coped with food stress by regularly reducing the number of meals eaten per day, as well as by rationing quantities of food eaten per meal.

Urban agriculture contributes significantly to urban food security in the period during which grain produced by UA lasts in a household. A total of

Table 4.5 *Comparing UA and food security*

Category	Food secure		Food insecure	
	UA participants *n = 87*	*Non-participants* *n = 5*	*UA participants* *n = 244*	*Non-participants* *n = 32*
Energy requirements met (%)	107	107.3	76	65
Food insecure (%)	26.3	13.5	73.7	86.5
Consumption coping				
Borrowed food	20.3	79.7	6.7	93.3
Eat less-preferred foods	25	74.4	10.5	89.5
Reduced number of meals	23	77	8.7	91.3
Reduced quantity per meal	24.1	75.9	9.1	90.9
Income coping				
Sold clothes	12.5	87.5	0	100
Accessing savings	23.4	76.6	14.3	85.7
More members seeking employment	17.5	82.5	15.8	84.2
Children in income-generating activities	16.1	83.9	25	75
Sub-letting assets	31.9	68.1	14.3	85.7
Expenditure coping				
Avoided expenditure on healthcare	19.8	80.2	0	100
Reduced expenditure on education	18.8	81.3	0	100
Reduced expenditure on water and electricity	23.6	76.4	11.1	89.9
Reduced expenditure on transport	21.8	78.2	0	100

Source: Survey data.

32.9 per cent of those households practising UA had enough grain to last one to three months, while 16.9 per cent of the households had enough grain for four to six months. Furthermore, close to 9 per cent of the households had enough grain to last seven to nine months and 14.8 per cent of the households had enough grain for 10–12 months. Roughly 27 per cent of the households had no grain at all, either because they did not produce maize that season or because the small amount they produced was consumed in less than a month.

Across all households surveyed, food is the largest item in the household budget (49 per cent), followed by transportation (35 per cent). Figure 4.1 shows that more than 70 per cent of total expenditure of households in the lowest expenditure quartile went to food, compared with less than 20 per cent in the highest quartile. The poor, who pay a disproportionate part of their income on food, end up being vulnerable to any unanticipated price changes or problems. This finding is consistent with the Engelian relationship between income and the percentage allocated to food (Colman and Young, 1996;

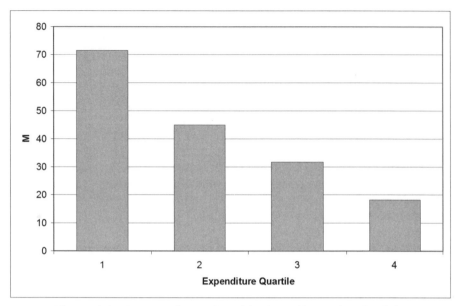

Figure 4.1 *Average share of total budget allocated to food, by expenditure quartile*

Deaton, 1997). As income increases, the percentage of the household budget allocated to food falls.

Harare urban diets were shown to be more varied than rural diets (ZIMVAC, 2003), with tubers and cereals dominating (62.2 per cent), followed by meat and fish (13.7 per cent), then by fats, oils and vegetables (15.7 per cent). Expenditure-reduction strategies included cutting spending on consumption and purchases mainly by curtailing money spent on transportation and health care to make ends meet.

Objective 4: Factors associated with food security and energy adequacy

The results of the energy adequacy regression using household calories per adult equivalence unit are given in Table 4.6. Adjusted R^2 was 0.487, meaning 48.7 per cent of the variation in energy adequacy is explained by the given independent variables. The Durbin-Watson statistic was 1.653. This suggests that there is a slight positive correlation between adjacent residuals, since the value is close to 2, signifying that the residuals are uncorrelated. This was also confirmed prior to moving on to regression analysis, where residuals were independent of each other.

Some variables showed a significant link with household energy adequacy. Practising UA was found to be related to energy adequacy. This is consistent with a number of earlier studies on UA (Mbiba, 1995; Mutangadura and Makaudze, 1999; Mwakiwa, 2004; Mudimu et al, 2005). Correlations also

Table 4.6 *Factors associated with adult diet and energy adequacy (linear regression model)*

	Non-standardized coefficients		Standardized coefficients	Student t	P-value
	β	Standard error	β		
(Constant)	79.825	8.533		9.354	**0.000***
Suburb type	−2.295	2.846	−0.046	−0.806	0.421
Sex of head	−1.190	3.181	−0.021	−0.374	0.708
Participation in UA	10.281	3.084	**0.131**	3.333	**0.002***
Head's contribution to UA	9.755	3.969	**0.133**	2.458	**0.015***
Health status of head	0.733	3.915	0.010	0.187	0.852
Plot ownership	0.223	2.968	0.004	0.075	0.940
House ownership	10.946	2.756	**0.234**	3.972	**0.001***
Household size	−4.036	0.615	**−0.378**	−6.561	**0.001***
Within-yard cultivation	0.744	2.756	0.016	0.270	0.787
Open-space cultivation	1.516	3.545	0.028	0.427	0.669
Plot cultivation	7.962	6.546	**0.077**	1.216	**0.020***
Informal employment	6.745	2.532	**0.140**	2.663	**0.008***
Formal employment	2.547	1.611	0.084	1.581	0.345

Adjusted R^2 = 0.487; Durbin-Watson = 1.653; *Significant at the 0.05 level.

were found between energy adequacy and household size, the head's contribution to UA labour, plot cultivation, per capita expenditure (which is being used as a proxy for income) and house ownership. Age and gender of the head of household and the type of suburb were not significantly associated with household energy adequacy.

As expected, households with more members were more likely to have lower energy adequacy values than those with fewer household members. This finding is consistent with ZIMVAC findings in both their urban and their rural assessments (ZIMVAC, 2003). Households whose heads contributed to UA labour were more likely to meet more of their energy requirements and be more food secure, according to the bivariate analyses already mentioned.

Urban agriculture taking place in spaces officially designated for UA – in contrast to in open spaces and small yards – was associated with a significantly greater energy adequacy. Therefore, households practising UA on plots are more likely to be food secure than those who are using open spaces and yards within residential areas.

Home owners had significantly higher energy adequacy compared with lodgers, leaseholders and those in accommodations provided by their company or employer. This might be because households that own their homes take much of the land on-plot if they are subletting from other renters and have greater claim to open spaces by virtue of long periods of residence in the suburbs (Mwakiwa, 2004).

Participation of household members in informal employment (vs the formal sector) was associated with greater energy adequacy. This probably reflects the fact that incomes from formal employment are much lower than returns from informal activities given the current status of the Zimbabwean economy.

Objective 5: Policies that enhance urban agriculture

Figure 4.2 is a representation of focus-group data and secondary data from the Municipal Development Partnership (MDP, 2003), the central government and Harare City Council. The views of both UA practitioners and policy makers from the central government and the municipal administration were included.

Urban agriculture practitioners' perspectives
→

		Positive	Negative
Policy makers' perspectives	**Positive**	**Q1** Enhancing food security Alleviating poverty Improving nutrition Generating additional income Creating employment	**Q2** Legislation that prohibits or restricts UA Lack of secure permanent tenure* Lack of integration of UA as an urban land use By-laws that discourage UA Closure of market stalls
	Negative	**Q3** Use of open spaces for cultivation Use of fertilizers Leaving stubble in fields Poor enforcement of by-laws Slow pace of policy implementation (regularization and control)	**Q4** Erosion Siltification Loss of biodiversity Habitat loss Pollution: water, soil, noise Deforestation Land shortage Diseases Increase in criminal activity Interference with traffic and some underground pipe networks Lack of support services (training and extension, credit) Lack of technologies appropriate for urban conditions Lack of organization among the farmers

* Policy makers consider this positive since it removes restrictions on their ability to change land uses and amend physical plans.

Source: MDP, 2003.

Figure 4.2 *Mapping of views of policy makers and practitioners on UA*

Their views of UA are categorized by the author as either positive or negative. Interestingly, some of the results were not as anticipated. The most predominant issues identified were included in the matrix. Those in Quadrant 1 are considered by all parties to be the positive contributions of UA. Quadrant 2 represents issues that the policy makers consider to be good, while practitioners view them as unconstructive and negatively affecting UA. Issues that are seen as harmful and negative by policy makers while they are viewed as helpful by the practitioners are given in Quadrant 3. Quadrant 4 represents issues identified by all parties as the problems of UA that negatively affect society, the environment or the planning process.

Quadrants 1 and 3 of Figure 4.2 represent issues that both groups agree upon, which can represent the basis for dialogue that seeks to ensure common agreement on UA. Quadrants 2 and 4 are boxes of conflict, divergent views and issues that need to be resolved. It is in these boxes where there is tension between the practitioners and the metropolitan and central government policy makers. If issues in these areas are resolved, there could be a more peaceful coexistence between the two parties. Policy should aim to lessen the rift and stand-off in these two quadrants by increasing dialogue.

From the above analysis, it can be concluded that most of the existing policies and legislation address the negatives listed in Quadrant 3, but not as many issues from Quadrants 1, 2 and 4 are being addressed. For example, for optimum management of land where investment is needed for such things as conservation works, people need land titles and security of tenure. While full tenure is unlikely to be feasible in the short term, one possible solution would be to provide secure yet informal tenure – such as by issuing temporary permits.

The Grain Marketing Board (GMB), which collects surplus grain from farmers for redistribution, was able to purchase more than 60 per cent of the annual national maize production in the early 1980s (Figure 4.3). This marks an improvement, because the GMB only managed to procure an average of 34 per cent per year during the 1970s. However, since 2000, the GMB was only able to collect 18 per cent of the total national production of maize per annum because of lack of surplus.

This poses a challenge regarding how to redistribute grain from surplus areas to urban centres. The situation necessitates the development of an integrated strategy of urban and rural development. Food security in Zimbabwe is achievable through a combination of urban production and improvements in the infrastructure for food distribution from rural areas.

Respectively, 62 and 42 per cent of respondents cited shortages of inputs as well as high costs as problems. This suggests the need for support for production inputs. For example, the city government can play an important role in providing enough market stalls from which to sell their crops. The imposition of generic and restrictive policies on UA has not succeeded. Harare City Council needs to replace these with policies that actively regulate and guide UA. If such policies were implemented, producers would be able to fully

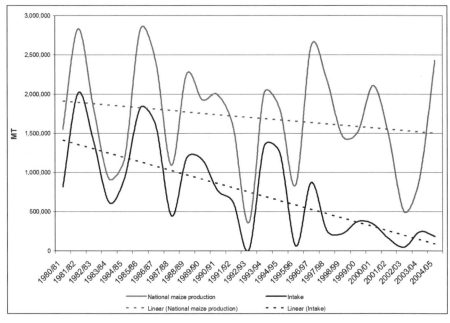

Source: GMB and CSO.

Figure 4.3 *Trends in maize intake via the Grain Marketing Board*

realize their potential and possible side-effects would be avoided (MDP, 2003). One option for Harare would be to create a platform on urban food and environmental policy issues that includes UA. In Bulawayo, the city has undergone a project called the Multi-stakeholder Process for Action Planning and Policy Design. Its purpose is to bring multiple stakeholders around the same table to discuss issues of common interest, challenges and solutions. A similar process for Harare City would improve the quality of decision making. The city can act as a convenor and facilitate the process of the development of UA. Such a municipal platform could lead to the development of a comprehensive, city-based food security plan as it has in Buluwayo. It could also stimulate the integration of UA into urban zoning and development plans, to advise on the revision of city by-laws related to UA, and to coordinate the planning of action programmes, as is the case in Kampala, Uganda.

CONCLUSIONS

As the population of Zimbabwe becomes increasingly urban, food insecurity becomes more of an urban issue. This study showed that UA can contribute significantly to food security. Therefore, policy guidelines – along with guidelines for sustainable use of open spaces for UA – should be both very clear and readily available. This is particularly relevant, given that 80 per cent of UA takes place in open spaces and on vacant public land. Also, it is important to support UA in the form of production inputs, agricultural extension, viable prices and places for marketing.

The regulatory environment should not constrain the activities of the urban self-employed, given that close to 50 per cent of the income of the lowest quartile income group comes from the informal sector. Historically, the municipality has tended to view the informal sector as a problem to be solved rather than as a viable resource for dealing with unemployment. Lately, the central government has also developed an intolerance for this sector, resulting in harassment of informal entrepreneurs. Female-headed households tended to have lower incomes because they had fewer assets, especially human assets such as education, labour and marketable skills. Therefore, investment in education for girls is likely to reduce poverty in the long term.

The municipality should invest in publicizing UA legislation and invest in dialogue with urban farmers. This is crucial, given that more than 40 per cent of the people practising UA were unfamiliar with any such related by-laws, and 22 per cent considered existing legislation to be hostile to UA. Given the potential environmental and health hazards that can emanate from UA, municipal administration should actively regulate, manage and guide UA. Only by doing this will they be able to fully realize its potentials and prevent negative side-effects.

ACKNOWLEDGEMENTS

I want to thank Canada's International Development Research Centre for fully funding my research project through its AGROPOLIS award programme. I am grateful to Agricultural Research and Extension Services (AREX) and the City of Harare for providing an environment conducive for my research. I also want to thank my supervisors, Dr R. Mano and Mrs C. Muchopa, and the many Harare residents who answered my questions.

NOTE

1   Charity Mutonodzo; MSc Agricultural Economics; 10 Hillside Gardens, 15–17 Ferreira Avenue, Hillside, Harare, Zimbabwe; tel: +263 4 747800; email: mutonhodzac@yahoo. co.uk

REFERENCES

Colman, D. and Young, T. (1996) *Principles of Agricultural Economics: Markets and Prices in Less Developed Countries*, Department of Agricultural Economics, University of Manchester, UK

Deaton, A. (1997) *The Analysis of Household Surveys: A Micro-econometric Approach to Development Policy*, World Bank, Washington, DC

FEWSNet (Famine Early Warning Systems Network) (2002) Emergency Food Security and Vulnerability Assessment Report, Washington, DC

FEWSNet (2005) *Approach to Livelihoods-Based Food Security Analysis Methodology*, FEWSNET, World Bank, Washington, USA

Food and Agriculture Organization (FAO) (1993) *Food and Nutrition in The Management of Group Feeding Programmes*. FAO, Rome, Italy

Maxwell, D., Levin, C., Amar-Klemesu, M., Morris, S., Ruel, M. and Ahiadekeke, C. (2000) 'Urban livelihood and food and nutrition security in greater Ghana', Research Report No 112, International Food Policy Research Institute, Washington, DC, USA

Mbiba, B. (1995) *Urban Agriculture in Zimbabwe: Implications for Urban Management and Poverty*, Avebury, Hants, UK

Moser, C. (1996) 'Confronting crisis: A comparative study of household responses to poverty and vulnerability', in *Four Poor Urban Communities*, World Bank, Environmentally Sustainable Development Studies and Monographs Series No 8, Washington, DC, USA

Mougeot, L. J. A. (1995) *Urban Food Production: Evolution, Official Support and Significance (with Special Reference to Africa)*, Urban Agriculture Notes City Farmer: Canada's office of Urban Agriculture; IDRC, Ottawa, ON, Canada

Mudimu, G., Matinhure, N., Mushayavanu, D., Chingarande, S., Toriro, P., Muchopa, C. (2005) *Research Project on Improving Access to Land for Urban Agriculture by the Urban Poor in Harare*, IDRC, University of Zimbabwe, Harare, Zimbabwe

Municipal Development Partnership (MDP) (2003) 'The Harare declaration on urban and peri-urban agriculture in Eastern and Southern Africa', www.mdpafrica.org.zw/Publications/Harare%20Declaration.pdf (accessed 15 September 2008)

Mutangadura, G. and Makaudze, E. (1999) 'Urban vulnerability to chronic poverty and income shocks and effectiveness of current social protection mechanisms: The case of Zimbabwe', Harare Consultancy Report (draft) submitted to the Ministry of Public Service, Labour and Social Welfare, and the World Bank, Harare, Zimbabwe

Mwakiwa, E. (2004) 'Evaluation of the socio-economic determinants and benefits of urban agriculture: the case of Kadoma, Zimbabwe', MSc Thesis, Agricultural Economics, University of Zimbabwe, Harare, Zimbabwe

Poate, C. D. and Daplyn, P. F. (1993) *Data for Agrarian Development*, Cambridge University Press, Cambridge, UK

Rabinovitch, J. (1999) 'Practical approaches to urban poverty reduction', paper presented by Urban Development Team at International Forum on Urban Poverty, Governance and Participation, Florence, UNDP, November 1999

Tabatabai, H. (1993) 'Poverty and food consumption in urban Zaire', Cornell Food and Nutrition Policy Programme Working Paper 46, Cornell Food and Nutrition Programme, Ithaca, NY, USA

United States Agency for International Development (USAID) (1996) 'Performance indicators for food security', Final output of the December 1995 USAID Workshop on Food Security Performance Measurements, USAID, CDIE, Washington, DC, USA

World Food Programme (WFP) (2000) *Food and Nutrition Handbook*, WFP, Rome, Italy

WFP (2005) *Emergency Food Security Assessment Handbook: Methodological Guidance for Better Assessments*, WFP, Rome, Italy

WFP/UN High Commission for Refugees (UNHCR) (1997) *Guidelines for Estimating Food and Nutritional Requirements*, WFP, Geneva, Switzerland

World Resource Institute (WRI) (1999) *1998–1999 World Resources: A Guide to the Global Environment*, WRI, Washington, DC, USA

Zimbabwe Vulnerability Assessment Committee (ZIMVAC) (2003) *Zimbabwe Urban Areas, Food Security and Vulnerability Assessment*, Harare, Zimbabwe

5

Urban Agriculture and Food Security in Lilongwe and Blantyre, Malawi

David Dalison Mkwambisi[1]

INTRODUCTION

Despite continued economic growth around the world, food insecurity remains a pressing problem in many parts of Africa (Garrett and Ruel, 1999; Maxwell, 1999; Mougeot, 2005a; UN-HABITAT, 2006). For example, the World Food Programme (FAO, 2001) and Food and Agriculture Organisation (FAO, 2003) both estimate that approximately 800 million people are unable to obtain an adequate and secure supply of food year round. The FAO (2002) further estimates that about 33 per cent of people in sub-Saharan Africa are undernourished. The UN-HABITAT (2006) reports that the percentage of urban residents in sub-Saharan Africa is expected to rise from 30 to 47 per cent of the total population. This will bring about new and very severe challenges for urban policy, especially when trying to ensure household food security (Huddad et al, 1998).

UA is one strategy where recent research suggests that food insecurity could be tackled (Pothukuchi and Kaufman, 1999; Mougeot, 2001, 2005a, b). UA has been shown to be an important source of food in developing countries and a critical food 'insurance policy' for poor urban households (Mougeot, 2000; Nugent, 2000). UA also affects household nutrition as it provides a source of fresh, locally grown crops that meets the micronutrient requirements in poor households' diets (FAO, 2001; Maxwell, 2001). Other benefits have been documented, such as increases in household incomes due to the sale of urban agricultural produce (Sanyal, 1985; Smit, 1996; Sabates et al, 2001; Henn, 2002; IFPRI, 2002).

Nowhere are these issues more pressing than in Malawi, where persistent poverty, HIV/AIDS and rapid urbanization have brought huge numbers of poor, unemployed and hungry people into the cities. Currently, over 55 per cent of the population is living on less than US$1 per day (Government of

Malawi, 2005a, b; USAID, 2005). Despite this, there is a real gap in Malawian policy, as UA is not seriously considered by the Malawian government as a viable livelihood option.

For example, despite UA being mentioned within the 'Town and Country Planning Act' (Government of Malawi, 1998a), there are no practical regulations to guide and support urban food production (Kwapata et al, 2001). The latest policy instruments put forward by the government still favour rural farmers but do not support or acknowledge urban food producers. As a result, the *Malawi Growth Strategy* (Government of Malawi, 2004), Malawi Poverty Reduction Strategy Paper (Government of Malawi, 2002) and previous models to reduce poverty in Malawi all fail to recognize the potential of UA.

The understanding of UA as a concept in Malawi is so limited that the mention of it is often based on speculation without any real awareness of who urban farmers are, what crops they produce, or the importance it plays in the local economy (Mkwambisi, 2005). Therefore, the purpose of this study was to evaluate the contribution UA makes to food security in Malawi. The information presented is designed to contribute to policy recommendations that are empirically grounded.

LITERATURE, CONCEPTS AND OBJECTIVES

The 1996 World Food Summit in Rome defined food security as a state when all people, at all times, have both the physical and economic access to sufficient, safe and nutritious food to meet their dietary needs and food preferences for an active and healthy life (FAO, 1996). The FAO (2002) explores this definition in more detail, pointing out that food security has two components: it is a phenomenon relating to individuals whereby the nutritional status of the individual household member is the focus; and food security should highlight the risk that this nutritional status may be undermined.

Garrett (2000) and the World Bank (1986) have developed similar themes by pointing out that urban food insecurity and malnutrition may be different from rural food insecurity because most urban dwellers depend on incomes to purchase their food. This relatively recent discussion hearkens back to Sen's (1991) approach that considered food security as a function of a person's – or household's – bundle of 'food entitlements'. According to this argument, people have different ways of meeting their household's food entitlements using the totality of rights and opportunities that they have (Sen, 1991).

Broadly speaking, Sen identified four types of entitlement: direct or production-based entitlement, which occurs when a person consumes or sells the food they produce; labour-based entitlement, which is obtained through working for a wage and purchasing food from the market; trade-based entitlements obtained through sale or barter of assets; and transfer-based entitlement, where entitlement is transferred through charity or food aid. Sen's work is important because it focuses on the ability of people to command food through legal means available to society (Sen, 1991). The entitlements framework puts the economic ability of individuals or families to deal with their own problems at the centre of the food security debate (Fraser et al, 2006).

The entitlement approach is useful for policy makers as it does not pre-suppose how people should be living, or the types of food they should be eating. Instead, it attempts to provide a framework that measures how much freedom an individual or a family has. Sen (2000) indicates that this freedom involves both the processes that allow freedom of actions and decisions, and the actual opportunities that people have, given their personal and social circumstances. Seen in this light, Sen's entitlement framework can help explore the complexity of UA. Food grown in the city can be a source of a range of different entitlements. It may provide a direct entitlement for those urban farmers who consume the food they produce. It can also provide two different types of indirect entitlement: first by providing marketable produce that a poor family could sell; and second by providing a source of paid employment for workers on larger-scale urban farms. Finally, if UA is used by charities, community- and faith-based organizations (CBOs and FBOs), it could also be used as the basis for transfer entitlements.

RESEARCH METHODOLOGY

Study location

Research was carried out on urban farms within the cities of Blantyre and Lilongwe in Malawi. They were useful for the purposes of this research because they are considerably different in terms of economic and demographic structure. Blantyre had a population of 711,233 in 2005, being the largest commercial and industrial centre in the country (Figure 5.1). It covers 228 km^2 of hilly ground, with a reasonably temperate climate, where 71 per cent of the city's residents live in unplanned settlements characterized by poor living conditions (Government of Malawi, 1998b).

Lilongwe lies on the Lilongwe-Kasungu Plain in the central fertile region of the country at an altitude of 1100 m above sea level. The capital of Malawi since 1975, Lilongwe is an administrative and commercial centre. Vast amounts of arable land and a low population density have resulted in an active UA and peri-urban agriculture (UPA) sector. According to the Government of Malawi (2005), Lilongwe has a population of 669,114 residents, where 40 per cent live below the poverty line. Within these two cities there is a range of low-, medium- and high-income neighbourhoods whose details are given in Table 5.1.

Data collection

Primary data was collected based on sustainable livelihood analysis through a structured questionnaire (Scoones, 1998; Ellis, 2000; Solesbury, 2003) that was administered to 330 heads of households who had access to agricultural land

Source: Government of Malawi, 2005a.

Figure 5.1 *Modified map of Malawi showing Blantyre and Lilongwe*

within the two cities. Specifically, 165 households were studied in Blantyre (112 male-headed and 53 female-headed) and 165 in Lilongwe (131 male-headed and 34 female-headed). The sample was stratified into high-income ($n = 70$: 68 males, two females) and low-income households ($n = 120$: 72 males, 48 females), as well as middle-income ($n = 140$: 127 males, 13 females) and also between male- ($n = 243$) and female-headed ($n = 87$) households. Due to several similarities in medium- and high-income households, results

Table 5.1 *Household characteristics in urban Malawi*

Character	Low income	High income
Literacy	High level of illiteracy	Low level of illiteracy
Population density	High	Medium to low
Settlement type	Unplanned	Well planned
Water source	Communal/unsafe	Indoor/safe
Main energy source	Fuel wood	Electricity
Waste collection services	Poor and insufficient	Excellent and timely
Criminal activities	High	Low
Food expenditure	57.5% of their income	29.8% of their income
Main daily income	K8.56	K55.57

Source: Government of Malawi (1998b).

have been presented for two groups only (low and high). Income status in urban Malawi was based on the government definition of poverty (Government of Malawi, 1998b).

Low- and high-income neighbourhoods were identified using consumption and expenditure data from the 1997–1998 Malawi Integrated Household Survey (Government of Malawi, 1998b), where a set of daily basic food and non-food requirements of individuals were identified. Respondents were selected using the snowball sampling method amongst low- and high-income households. Within each neighbourhood, key informants were identified through community workshops. Each respondent was then asked to identify two other urban farmers. This process was continued until 165 interviews were completed in each city.

The questionnaire used the 'five practical methods' outlined by Neefjes (2000). This method proposes that data be collected on basic household demographic characteristics such as marital and socio-economic status; income and employment; food frequency, allocation and food-related coping strategies; and urban agronomical practices. Logistic regression analysis and *t*-tests for independent samples were used to determine which groups (income, gender, education or location) benefited the most in terms of food security from UA.

Determining urban agriculture's contribution to food security

To assess UA's contribution to direct entitlements, respondents were asked to calculate the amount of food consumed that came from their own urban agricultural plots. Specifically, this research analysed the amount of maize each household produced and converted non-maize crops into 'cereal equivalents'. For example, fresh cassava and sweet potato yields (in kilograms per hectare) were considered to be worth 25 per cent of the equivalent fresh weight in cereal (FAO/WFP/GoM, 2005).

Results were then compared with the Government of Malawi's recommended annual consumption levels of 181 kg of cereal per capita.[2] Finally, to facilitate comparison with other entitlement strategies, data on food consumption were converted into a monetary value by using the 2005 market prices for maize in May of 2005 (21 Malawian kwacha/kg).

RESULTS

Results of the study revealed that the majority of households in Malawi were not dependent on UA as the main household food entitlement.

The findings from our study contrasted with other research conducted elsewhere in Africa which suggested that urban farming was an important source of food for urban populations (Maxwell, 1995). Our study showed that 70 per cent of all households gave formal employment as the most important component in their livelihoods' strategies, informal employment came second, while UA was ranked third.

Table 5.2 *Main livelihoods' sources as identified by household heads in urban Malawi*

Description	n	UA	Rural agriculture	Business	Formal employment	Informal employment
Lilongwe	165	9.7	4.8	17.6	66.7	1.2
Blantyre	165	24.8	4.8	13.9	53.9	2.4
Low-income households	120	42.5	2.5	25.8	25.8	3.3
High-income households	210	2.9	6.2	10.0	80.0	1.0
Female-headed households	87	55.2	3.4	17.2	24.1	0.0
Male-headed households	243	3.7	5.3	15.2	73.3	2.5
All households	330	17.3	4.8	15.8	60.3	1.8

Despite this low rating of UA, considerable variation between groups was observed. For example, while UA contributed 9 per cent of high-income households' livelihoods, it provided 42.5 per cent of low-income households. Thus, low- and high-income households are statistically different ($P < 0.05$), as shown in Table 5.2.

Results presented in Figure 5.2 show that high-income households were more productive in terms of the amount of food they produce than lower-income

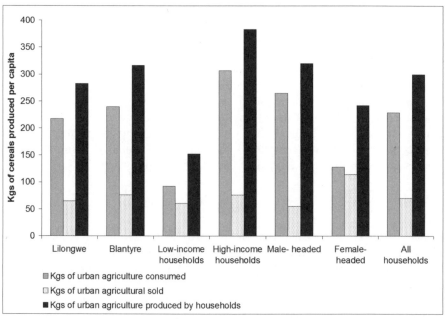

Figure 5.2 *Average cereal equivalents (expressed in kilograms per member of each household per year) from UA*

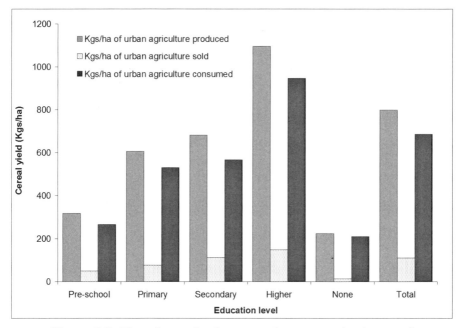

Figure 5.3 *The relationship between education and UA to total household's food bundle*

groups. This is due to the fact that higher-income groups had more access to larger plots and inputs such as fertilizer.

Female-headed households sold more of their crops as compared to other groups. Similar findings in Kampala were reported by Maxwell (1995). Logistic regression showed that there was statistical difference ($P < 0.05$) between income groups and gender but not between locations. Households headed by individuals with higher education recorded higher cereal yields than those who had not attended formal school (Figure 5.3).

Based on level of education, independent sample tests revealed a statistically significant difference between household heads with pre-school, no education and higher education ($P < 0.05$) on total yield and cereals consumed. However, there was no statistically significant difference between primary and secondary levels, secondary and higher levels, or illiterate and pre-school education levels. Also, there was no significant difference between education levels on cereals sold. The list of constraints farmers experienced are presented in Figure 5.4, which shows that female (49.4 per cent), low-income households (62.5 per cent) and farmers from Blantyre (53.9 per cent) mentioned agricultural land as the major constraint.

The average plot size available to male-headed households was 0.24 ha, while females had 0.08 ha, and the high-income households had an average of 0.27 ha as compared to 0.06 ha for low-income households. Lilongwe residents had on average 0.22 ha, while those from Blantyre had 0.17 ha per household.

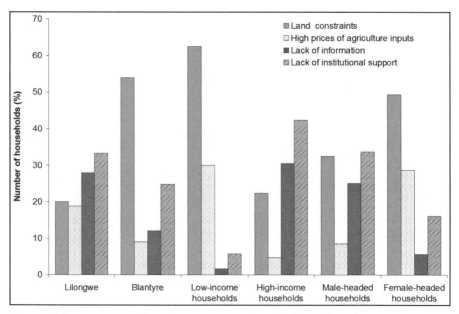

Figure 5.4 *Main agricultural constraints identified by urban farmers (n = 330)*

One interesting result was that, at small-farm sizes, both low- and high-income urban farmers obtain similar yields per hectare. However, for larger farms, high-income households are able to capitalize on economies of scale and become more efficient (Figure 5.5). This suggests that, even if land is

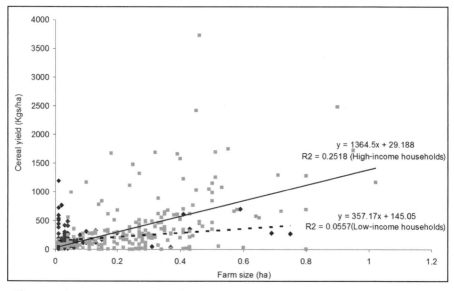

Figure 5.5 *Average cereal yield (kilograms per hectare) between low-income (n = 120) and high-income (n = 210) households in urban Malawi*

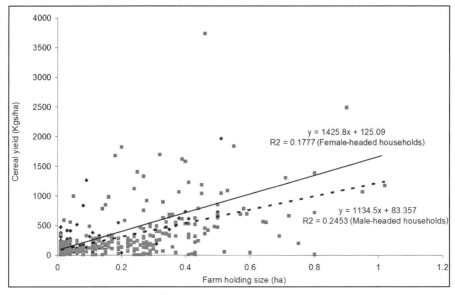

Figure 5.6 *Average cereal yield (kilograms per hectare) between female-headed (n = 87) and male-headed (n = 243) households in urban Malawi*

made available to the land resource poor farmers, the yield will remain low due to other production constraints such as inputs and access to information and technology. Overall, however, yields on UA plots remained low: the government suggests that farmers should obtain 6000–8000 kg/ha (Government of Malawi, 1999), and this reveals that even at its most productive, UA under-performs.

Despite the fact that women on average were poorer and less educated than their male counterparts, when they accessed large plots they were more efficient farmers because they had more time to attend to the crops (Figure 5.6). Thus, there were significant differences $(P < 0.05)$ on the cereal yields between male- and female-headed households.

DISCUSSION

Income status and urban food production

The study revealed that UA in Malawian cities will not provide the solution to food insecurity because it provides more food for middle- or upper-class families than for the poor. These results have been reported in other studies (Machethe et al, 1997; Maxwell, 1999). However, UA is playing a crucial role in generating extra income to some of the marginalized groups, especially the low-income, female-headed households and those with no formal education.

Gender and urban agriculture

According to Hovorka (2005), women have limited access to agricultural land and only 4.5 per cent participated in UA. When provided with adequate land, however, women were more efficient producers than men, which agrees with Scanlan (2004), who commented that women's role in agriculture covers all the production stages, which include acquisition, processing and preparation. UA in Malawi is dominated by men and high-income people who are able to invest and undertake UA as a 'luxury' livelihood strategy, and are able to access information, private and expensive agricultural consultants/experts and agro-input.

Urban agriculture and education

Education is known to be a major determinant of living standards, and information on education and literacy status is essential for planning and evaluation of existing policies; and statistics for Malawi showed that the majority of poor farmers were illiterate (World Bank, 1986; Government of Malawi, 2005a). Correlation between education and food security reveals that another challenge for the policy makers to surmount before the country can achieve sustainable UA is literacy: the majority of people in Malawi remain illiterate.

Production constraints

Different social groups cited different constraints including access to land, poor product presentation, unstructured markets, lack of government support and shortage of extension services (see also Maxwell, 1995). UA was one of the livelihood strategies for the many low-income households, but street vending and small-medium enterprises are preferred to farming. This observation confirms Ellis's (2000) comment that diversification into different sources of income has been used as a mechanism by poor households, especially in the informal market, and should be taken into consideration in urban development issues (Mougeot, 2005b). An increased consumption of street food has negative consequences on the nutrition and health status of the people due to a shift in dietary patterns (Ruel et al, 1999; quoted by Kennedy, 2003).

CONCLUSION AND POLICY RECOMMENDATIONS

Urban residents in Malawi engaged in agricultural activities mainly for consumption irrespective of their socio-economic status. However, many households did not consider urban farming as a commercial entity due to lack of support from the government and other institutions. The study has shown that it represents a safety net to marginal female-headed and low-income households. If the development strategy remains as it is in rural Malawi (to provide improved living conditions to the poor and disadvantaged societies), then government policies should encourage collective groups that would practise farming as a business.

ACKNOWLEDGEMENTS

The author wishes to express sincere gratitude to the International Development Research Centre (IDRC) through AGROPOLIS for funding the research work. Thanks also to the Country Director and Staff of Africare, along with research assistants and technical staff at Bunda College of Agriculture, for their contributions and considerable knowledge.

NOTES

1   David Dalison Mkwambisi; Ph.D. Environment and Development, University of Leeds, The School of Earth and Environment, Sustainability Research Institute, Leeds LS2 9JT, UK; tel: +44 113 343 6466; fax: +44 113 343 6716; email: Mkwambisi@env.leeds.ac.uk
2   This is made up of 150 kg maize and 13 kg of rice, sorghum, millet and wheat, and an equivalent of 17-kg-worth of grain from other sources such as sweet potato – crops such as sweet potato are considered to be worth 25 per cent of the equivalent amount of grain (FAO/WFP/GoM, 2005).

REFERENCES

Ellis, F. (2000) *Rural Livelihoods and Diversity in Developing Countries,* Oxford University Press, Oxford, UK
FAO (1996) 'Rome declaration on world food security and world food summit plan of action', World Food Summit, 13–17 November 1996, Rome, Italy
FAO (2001) *Urban Agriculture and Peri-urban Agriculture: A Briefing Guide for Successful Implementation of Urban Agriculture and Peri-urban Agriculture in and Countries of Transition,* FAO, Rome, Italy
FAO (2002) 'Food security and trade: An overview', in *Trade and Food security: Conceptualising the Linkages Expert Consultation,* Rome, 11–12 July 2002, FAO, Rome, Italy, www.fao.org/DOCREP/005/Y4671E/y4671e05.htm (accessed 24 September 2006)
FAO (2003) *The state of food insecurity in the world. Monitoring progress towards the World Food Summit and Millennium Development Goals,* FAO, Rome, Italy
FAO/WFP/GoM (2005) 'Crop and food supply assessment mission to Malawi', Special report by FAO/WFP mission to Malawi, July 2004, Ministry of Agriculture, Lilongwe and FAO, Rome, Italy, www.fao.org/docrep/008/J5509e/J5509e00.htm (accessed 1 September 2006)
Fraser, E. D. G., Hubacek, K. and Korytarova, K. (2006) 'Food and the environment: (Neo-) Malthusianism arguments and population growth', in O'Hara, P. (ed) *International Encyclopaedia of Public Policy, Governance in a Global Age,* Cambridge University Press, London
Garrett, J. L. (2000) 'Achieving urban food and nutrition security in the developing world: A 2020 vision for food, agriculture and the environment', Focus 3, Brief 1 of 10, www.ifpri.org/2020/focus/focus03/focus03.pdf (accessed 24 September 2006)
Garrett, J. L. and Ruel, M. T. (1999) 'Are determinants of rural and urban food security and nutrition status different? Some insights from Mozambique', *World Development,* vol 27, no 11, pp1955–1975
Government of Malawi (1998a) *Town and Planning Act,* Ministry of Local Government, Lilongwe, Malawi

Government of Malawi (1998b) *Integrated Household Survey I: Household Socio-Economic Characteristics*, National Statistics Office, Zomba, Malawi

Government of Malawi (1999) *Guide to Agricultural Production*, Agricultural Extension Services Department, Ministry of Agriculture, Lilongwe, Malawi

Government of Malawi (2002) *Malawi Poverty Reduction Strategy Paper*, Government of Malawi/IMF/World Bank, Lilongwe, Malawi

Government of Malawi (2004) *Malawi Growth Strategy*, vol III, Ministry of Economic Planning and Development, Lilongwe, Malawi

Government of Malawi (2005a) *Integrated Household Survey II (2004–2005)*, vol 1, *Household Socio-Economic Characteristics*, National Statistics Office, Zomba, Malawi

Government of Malawi (2005b) Malawi Food Security and Nutrition Policy, Government of Malawi, Ministry of Agriculture, Lilongwe, Malawi

Henn, P. (2002) 'User benefits of urban agriculture in Havana, Cuba: An application of the contingent valuation method', City Farmer, Vancouver, Canada, www.cityfarmer.org/havanabenefit.html (accessed 23 September 2006)

Hovorka, A. (2005) 'Gender, commercial urban agriculture and urban food supply in Greater Gaborone, Botswana', in Mougeot, L. J. A (ed) *AGROPOLIS: The social, political and environmental dimensions of urban agriculture*, International Development Research Council, Ontario, Canada, pp137–152

Huddad, L., Ruel, M. and Garrett, J. (1998) *Growing Urban Poverty and Undernutrition and some Urban Facts of Life: Implications for Research and Policy.* IFPRI, Washington DC, USA

IFPRI (2002) 'The Accra urban food and nutrition study', IFFPRI Issue Brief No 9, Addendum, IFPRI, Washington DC, USA

Kennedy, G. (2003) 'Food security in the context of urban Sub-Saharan Africa', http://foodafrica.nri.org/urbanisation/urbspapers/GinaKennedyFoodsecurity.pdf (accessed 23 September 2006)

Kwapata, M., Nkanaunena, G. A., Maliro, M. F. A., Kamwendo, P. M., Dawa, G. and Bokosi, J. (2001) *The Current Status and Contribution of Urban Agriculture and Identification of Information, Communication and Training Needs in Malawi*, Bunda College, Lilongwe, Malawi

Machethe, C. L., Reardon, T. and Mead, D. C. (1997) 'Promoting farm/non-farm linkages for employment of the poor in South Africa: A research agenda focused on small-scale farms and agro industry', *Development Southern Africa*, vol 14, no 3, pp377–394

Maxwell, D. G. (1995) 'Alternative food security strategy: A household analysis of urban agriculture in Kampala', *World Development*, vol 23, no 10, pp1669–1681

Maxwell, D. G. (1999) 'The political economy of urban food security in Sub-Saharan Africa', *World Development*, vol 27, no 11, pp1939–1953

Maxwell, D. G. (2001) The importance of urban agriculture to food and nutrition. Annotated biography, ETC-RUAF, CTA publishers, Leusden, Netherlands

Mkwambisi, D. D. (2005) 'Impact of urban agriculture on child labour, health and education in Lilongwe city, Malawi', in Zeleza Mtafu Manda (ed) *Malawi Physical Planning*, Alma Publishers, Mzuzu, Malawi, pp 94–111

Mougeot, L. J. A. (2000) 'Achieving urban food and nutrition security in developing countries: The hidden significance of urban agriculture', IFPRI, Brief paper number 6, 2000 www.ifpri.org/2020/focus/focus03/focus03.pdf (accessed 23 September 2006)

Mougeot, L. J. A. (2001) 'Urban agriculture: Definition, presence, potential and risks', in Barker, N., Dubbelling, M., Gindel, S., Sabel-Koschella, U. and de Zeeuw, H. (eds) *Growing Cities, Growing Food – Urban Agriculture on the Policy Agenda*, DSE, Eurasburg, Germany, pp1–42

Mougeot, L. J. A. (2005a) Introduction, in Mougeot, L. J. A. (ed) *AGROPOLIS: The Social, Political and Environmental Dimensions of Urban Agriculture*, International Development Research Council, Ontario, Canada, pp1–29

Mougeot, L. J. A. (2005b) 'Neglected issues on form and substance of research on urban agriculture', in Mougeot, L. J. A. (ed) *AGROPOLIS: The Social, Political and Environmental Dimensions of Urban Agriculture*. International Development Research Council, Ontario, Canada, pp267–279

Neefjes, K. (2000) *Environments and Livelihoods: Strategies for Sustainability*, Oxfam, London

Nugent, R. (2000) 'The impact of urban agriculture on the household and local economies', in Barker, N., Dubbelling, M., Gindel, S., Sabel-Koschella, U. and de Zeeuw, H. (eds) *Growing Cities, Growing Food-Urban Agriculture on the Policy Agenda*, DSE, Eurasburg, Germany, pp67–97

Pothukuchi, K. and Kaufman, J. L. (1999) 'Placing the food system on the urban agenda: The role of municipal institutions', *Agriculture and Human Values*, vol 16, no 2, pp213–224

Sabates, R., Gould, B. W. and Villarreal, H. J. (2001) 'Household composition and food expenditures: a cross-country comparison', *Food Policy*, vol 26, pp571–586

Sanyal, B. (1985) 'Urban agriculture: Who cultivates and why?', *Food and Nutrition Bulletin*, vol 7, no 3, pp15–25

Scanlan, S. J. (2004) 'Women, food security and development in less-industrialized societies: Contributions and challenges for the new century', *World Development*, vol 32, no 11, pp1807–1829

Scoones, I. (1998) 'Sustainable rural livelihoods – A framework for analysis', University of Sussex, Institute for Development Studies, Brighton, UK

Sen, A. K. (1991) *Poverty and Famines: An Essay on Entitlements and Deprivation*, Oxford University Press, Oxford, UK

Sen, A. K. (2000) *Development as Freedom*, 1st edition, Anchor Books, New York, NY, USA

Smit, J. (1996) *Urban agriculture – Food Jobs and Sustainable Cities*, United Nations Development Programme, New York, USA

Solesbury, W. (2003) 'Sustainable livelihoods: A case study of the evolution of DFID Policy', DFID Working Paper No 217, www.odi.org.uk/publications/working_papers/wp217.pdf (accessed 1 October 2006)

UN-HABITAT (2006) *The State of the World's Cities, 2001*, United Nations Center for Human Settlements, Nairobi, Kenya

USAID (2005) *Facts about Malawi*, USAID, Lilongwe, Malawi

World Bank (1986) *Poverty and Hunger: Issues and Options for Food Security in Developing Countries*, World Bank Policy Study, World Bank, Washington DC, USA

6

An Analysis of the Quality of Wastewater Used to Irrigate Vegetables in Accra, Kumasi and Tamale, Ghana

Philip Amoah[1]

INTRODUCTION

Cities in developing countries, including Ghana, are experiencing unparalleled population growth. Rapidly increasing water supply and sanitation coverage generates large volumes of wastewater, which is often released untreated into the environment (streams, drains, etc.). In Ghana, most urban centres have no means of treating wastewater and the sewerage network serves a small percentage (4.5 per cent) of the total population (GSS, 2002). The need for year-round production of vegetables in or near urban areas makes irrigation necessary; hence, farmers in search of water for irrigation often rely on wastewater. However, despite its importance for countless livelihoods, the use of urban wastewater for irrigation is not without health risks. Wastewater contains the full spectrum of pathogens found in the urban population, many of which can survive for several weeks when discharged onto fields.

Effective wastewater treatment can reduce pathogen levels, but in most developing countries it is not an option for the municipal authorities due to the high costs involved (Keraita et al, 2002). Most new sewerage treatment plants in Ghana are also operating below their design capacity. As wastewater treatment does not appear to be a realistic option, banning the use of polluted water by urban farmers has also been tried in Accra and other cities. This has failed because such bans threaten the livelihoods of many individuals, which run contrary to poverty-alleviation strategies. In these circumstances, urban farmers express significant concerns because their livelihoods are at permanent risk. Any solution to reduce health risks without forcing them to change their (market-driven) cropping patterns or access to water would be appreciated. In addition, Ghana's Tourism Board has started a campaign directed at consumers to promote 'safer vegetables for healthier cities'. This was prompted because tourists were suffering from outbreaks of gastrointestinal disorders after consuming vegetables in urban areas.

Another potential health risk derives from the use of pesticides, although this is beneficial in decreasing crop loss both before and after harvest (Clarke et al, 1997). Despite the recognition of urban agriculture (UA) as a source of urban food security, concerns are growing among city authorities on the indiscriminate use of pesticides. Insufficient data exist, however, on the actual gravity of the problem. These would provide guidance on appropriate interventions or policy formulation.

Hence research into other risk-reduction options is required for regions or countries where wastewater treatment is not a realistic option. Therefore, this study was designed to explore the contamination and decontamination of wastewater-irrigated crops. The structure of this chapter first outlines the concepts, objectives and hypotheses, then describes the research methods followed by results, a discussion of impacts and finally some recommendations.

KEY CONCEPTS

Wastewater

Wastewater is the liquid portion of waste. It may be defined as a combination of liquid or water-carried wastes that are removed from residences and institutions, as well as commercial and industrial establishments. In addition, a combination of groundwater, surface water and storm water may be present (Metcalf and Eddy, 1995). In this section, it is assumed that urban wastewater may be a combination of some or all of the following:

* *Municipal wastewater* consists of domestic effluent made up of black water (excreta, urine and associated sludge) and grey water (kitchen and bathroom wastewater). It may also include water from commercial establishments and institutions, including hospitals.
* *Industrial effluent* is water polluted by industrial processes and containing high levels of heavy metals or other chemical or organic constituents. Industrial effluent does not normally contain high levels of microbiological pollution unless it emanates from slaughterhouses or food-processing plants.
* *Storm water* is run-off precipitation that finds its way across surfaces into receiving waters. Urban storm run-off is collected and transported in storm or combined sewers. The composition of storm water reflects the composition of precipitation and the surfaces with which it is in contact (Environment Canada, 2006).

Other terms

* *Marginal-quality water:* water whose quality might pose a threat to sustainable agriculture and/or human health, but which can be used safely

for irrigation provided certain precautions are taken (Abbot and Hasnip, 1997). Such water is polluted as a consequence of mixing with wastewater or agricultural drainage (Cornish et al, 1999).

•   *Indirect use of wastewater:* This is the unplanned application to land of wastewater from a receiving water body. Municipal and industrial wastewater is discharged without treatment or monitoring into the watercourses draining an urban area. Irrigation water is drawn from rivers and streams or other natural water bodies that receive wastewater flows. There is often no control over the use of water for irrigation or domestic consumption downstream of the urban centre. Consequently, many farmers indirectly use marginal-quality water of unknown composition that they draw from many points downstream of the urban centre.

Research objectives

This chapter has four primary objectives:

1   To assess the water quality (biological and chemical) of irrigation water sources used for vegetable cultivation.
2   To trace the (microbiological and helminth) contamination pathways of vegetables in urban and peri-urban sites to identify where interventions should take place along the production–consumption continuum.
3   To isolate and identify fecal coliform (FC) bacteria found on irrigated vegetables from urban and peri-urban sites.
4   To determine the level of pathogen and pesticide contamination of vegetables produced on urban agricultural sites.

Hypotheses

Three basic hypotheses underlie this research:

1   Fecal coliform (FC) and helminth egg population levels in wastewater from different urban sources exceed common standards recommended for irrigation.
2   Microbiological (FC and helminth levels) contamination of wastewater-irrigated vegetables is increased through handling and distribution within the production marketing chain.
3   Potential health risks to consumers are not reduced to acceptable levels after the normal household treatment of vegetables.

METHODS

Phases of research and cities studied

The study was divided into three phases. The first phase (market sampling) was conducted in the three Ghanaian cities of Accra, Kumasi and Tamale (Figure 6.1). The second phase (water to field to market sampling) took place

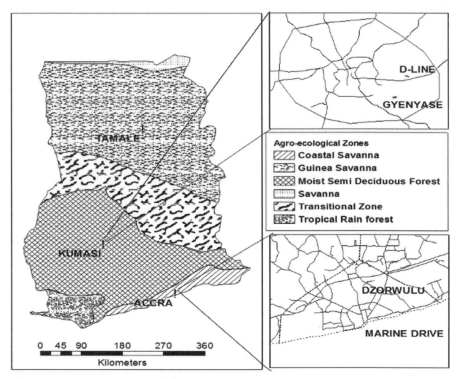

Figure 6.1 *Agro-ecological map of Ghana showing the three urban study areas of Tamale, Kumasi and Accra, with details for Kumasi and Accra*

in Accra and Kumasi. The third phase (field trials) was only conducted in Accra.

Accra is the capital city of Ghana, with a population of approximately 1.7 million (GSS, 2002). It is located at the Gulf of Guinea in the coastal savannah belt. Two sites were chosen. Dzorwulu is one of Accra's suburbs, with a major vegetable production site within the metropolis. It has a total land area of roughly 12 ha, which is cultivated by more than 300 farmers. Marine Drive is a smaller vegetable production site in Accra near Independence Square. Its area is close to 4 ha and has more than 100 vegetable growers.

Kumasi is the capital town of the Ashanti Region and the second largest city in Ghana, with a population which is a little more than one million (GSS, 2002). It is located in the forest belt of Ghana. Two sites were chosen here, also. The Gyenyase site is the largest urban vegetable growing site in Kumasi, with a total land area of approximately 6 ha which supports approximately 60 vegetable farmers. D Line, a suburb close to the Kwame Nkrumah University of Science and Technology (KNUST), is a popular vegetable farming spot in Kumasi, with close to 40 farmers and a total area of approximately 3 ha.

Tamale is the administrative and regional capital of Ghana's northern region, located in the savannah zone with a population of approximately

300,000 (GSS, 2002). In contrast to Accra and Kumasi, Tamale Municipality has few bodies of water and only a few seasonal streams.

Phase I: Sampling of vegetables at the markets

The purpose of this phase was to determine pathogens and pesticide contamination of vegetables produced at UA sites. From October to December 2002, a total of 180 vegetable samples (lettuce, cabbage and spring onion) were collected from nine major markets and 12 specialized, individual vegetable and fruit sellers (i.e. sellers with permanent stalls outside of designated markets) in Accra, Kumasi and Tamale (see Figure 6.1). At each market, samples were collected under normal purchase conditions from three randomly selected sellers. A minimum of three composite samples – each containing two whole lettuce heads, three bunches of spring onions (each containing two bulbs) and three cabbages – were collected from the upper, middle and lower shelves of each seller, put in sterile polythene bags and transported on ice to the laboratory, where they were analysed immediately or stored at $4°C$ until analysis could occur within 24 hours.

These and all other samples collected (see below) were analysed for coliform and helminth egg populations using the most probable number (MPN) method (APHA/AWWA/WEF, 2001) and the floatation and sedimentation method following a modified US Environmental Protection Agency (US–EPA) method (Schwartzbrod, 1998), respectively. Gas Chromatography (Flame Ionization Detector; Hewlett-Packard 5890 series II) was used for pesticide residues on lettuce following the method adopted by Ntow (2001). Sample peaks were identified by their retention times compared with those of the corresponding pesticide standard obtained from the International Atomic Energy Agency. The ability of the laboratory to identify these substances has been verified by cross-tests of river sediments in Ghana.

Phase II: Contamination pathway study

This study was undertaken to determine the microbiological contamination levels at various entry points along the production-marketing chain. In Accra and Kumasi, two major irrigated vegetable production sites were selected based upon the source of irrigation water and the type of vegetables grown, with emphasis on exotic vegetables such as lettuce that were probably going to be consumed raw. Both sites in Accra used water from drains and streams, while shallow wells and streams provided the sources in Kumasi. Farmers in at least one of the two sites in each city used irrigated, piped water as their source over a period of at least three years. All sites had a similar history of land use. For instance, all were under vegetable cultivation for periods of not less than five years, and all farmers used poultry manure as a source of fertilizer.

Irrigation water

This study monitored the microbiological quality of irrigation water from different urban sources. One composite sample per week was collected from each source for 52 weeks from May 2003 to April 2004. In all, six were involved – stream, shallow well and piped irrigation water sources in Kumasi and drain, stream and piped water in Accra. Sampling at all sites was carried out between eight and ten in the morning in keeping with farmer's irrigation practices (APHA/AWWA/WEF, 2001). At each site, 200-ml glass bottles were used to take water from three different points in the wells or in 20-m intervals along the drain or stream. Piped water was collected directly from the water hose used by the farmers for irrigation. Samples from a particular site were later joined into one composite sample per source and transported to the laboratory on ice. A total of 312 composite water samples were analysed for total and fecal coliform populations. Sampling for helminth egg quantification in irrigation water was done twice every month for five months from November 2003 to March 2004 at all the selected sites. Two-litre samples were taken after deliberately disturbing the bed of the irrigation water source to stimulate agitation that might occur when farmers are filling their watering cans. This was intended to bring out the eggs, as they usually settle under their own weight (Cornish et al, 1999).

Lettuce sampling in farms, wholesale and retail markets

Over a period of 12 months, from May 2003 to April 2004, a total of 1296 lettuce samples were collected at different entry points[2] from farm to the final retail outlet. The original sets of lettuce were either irrigated with stream, drain, well or piped water (microbiological quality of these water sources were monitored as described above). Twice a month, a minimum of three composite samples (each containing two whole lettuces) from each of the selected farm sites were randomly collected using sterile disposable gloves just before harvesting for sale at the market. These were put into separate sterile polythene bags and labelled as farm samples. The seller was followed to the wholesale market where another sample from the same original stock was collected, before being finally sold to a retailer. At the final retail point, three composite samples were again sampled after vegetables were displayed on the shelves for at least two to three hours, which is a typical turn-over period at the retail point. Producers and sellers were paid for their produce. Sampled vegetables were transported on ice to the laboratory where they were analysed for TC, FC and helminth counts. To eliminate potential biases during analysis, staff working in the laboratories were blind to the source of the samples.

Phase III: Field trials

In order to further understand the importance of the different possible sources of contamination of irrigated lettuce (e.g. wastewater, poultry manure, soil), a field trial was set up at two farm sites in Accra. Lettuces were grown on raised beds of the common size of approximately $3\,m^2$. For each trial two adjacent plots, A and B, were used, with each plot subdivided into four blocks. Each block contained four beds, making a total of 16 beds per plot. Using a randomized complete block design, each of the four beds in a block was randomly allocated to one of four treatments (three representative poultry manure samples, PM_1, PM_2 and PM_3, with average FC populations of 4.3×10^7, 2.4×10^5 and 3.3×10^3 $100\,g^{-1}$, respectively, and one inorganic fertilizer NPK (F)), which were applied to the beds two weeks after transplanting following farmers' usual practices. Each of the two plots (A and B) at a site was irrigated with either clean piped water or wastewater from a drain. The crops on a bed were irrigated once daily with approximately $30\,l$ of water every morning excluding rainy days. All work was done by the farmers as part of their standard routine.

Soil samples from the top $10\,cm$ were collected from six randomly selected points on each bed, just before transplanting from the nursery, and combined into one composite sample per bed for the initial count of FC and helminth egg populations in duplicates. Samples were also taken from nearby plots with no previous history of vegetable production to serve as controls. Six lettuce heads were aseptically collected from each bed four weeks after transplanting, after which they were randomly combined into three composite samples containing two lettuce heads. The samples were kept in a refrigerator at $4°C$ and analysed within 24 hours for FC and helminth eggs using standard methods.

Data analysis
The data were analysed using SPSS for Windows 10. Using MPN, FC populations were normalized by log transformation for the analysis of variance (ANOVA). *t*-tests, used to compare data sets, were also employed where necessary. Inferences for significant differences were set at $P < 0.05$.

Results and Discussion

Microbiological quality of market vegetables from urban markets
Bacterial contamination of market vegetables
Most of the vegetable samples showed high FC contamination levels (see Table 6.1). The highest level of FC contamination was recorded in lettuce (geometric mean count of 1.1×10^7 per g wet weight), probably due to the

Table 6.1 *Ranges of total and fecal coliform population on selected vegetables*

Vegetable	MPN g^{-1} wet weight	
	Total coliform	Fecal coliform
Lettuce	9.3×10^5 to 1.5×10^{11}	4.0×10^3 to 9.3×10^8
Cabbage	2.6×10^5 to 1.5×10^{11}	1.4×10^4 to 2.8×10^7
Spring onion	9.3×10^5 to 1.9×10^{10}	1.5×10^4 to 4.6×10^8

Note: MPN, most probable number.

larger surface area exposed. Cabbage and spring onion showed geometric mean counts of $3.3 \times 10^6\,g^{-1}$ and $1.1 \times 10^6\,g^{-1}$ wet weight, respectively. No sample had less than 4000 FC per gram wet weight.

The mean FC levels of all the three crops exceed the International Commission on Microbiological Specifications for Food (ICSMF, 1974) recommended level of 10^3 fecal coliforms per gram fresh weight. Several factors may account for the high levels recorded in most of the analysed vegetables. Among these is the use of polluted irrigation water and fresh poultry manure, both of which are applied on top of the crops. Significantly high FC contamination levels (between 4.8×10^3 and 2.8×10^6 $100\,ml^{-1}$) which usually exceed common standards have been recorded in irrigation water (Cornish et al, 1999; Drechsel et al, 2000; Mensah et al, 2001; Keraita et al, 2002). High FC populations (between 3.6×10^4 and 1.1×10^7) were also reported in poultry manure in the same study areas.

Another potential source of contamination is market-related handling, especially where provision for better sanitary standards (e.g. clean water for crop washing and refreshing) is lacking. A relatively high total and FC population recorded on some vegetables was also reported by Johnson (2002) and Armar-Klemesu et al (1998) analysing street food and market crops in Accra, respectively.

Mean helminth egg population on vegetables

About 30 per cent of vegetables had no helminth eggs. Lettuce, cabbage and spring onion carried mean helminth egg populations of $1.1\,g^{-1}$, $0.4\,g^{-1}$ and $2.7\,g^{-1}$ wet weight, respectively. No significant difference was observed in the mean helminth egg populations recorded in lettuce and cabbage; however, the difference between spring onion and both lettuce and cabbage was significant ($P < 0.05$). If distributions skewed, the mean separations were compared using one-way ANOVA. The eggs identified included *Ascaris lumbricoides, Ancylostoma duodenale, Schistosoma heamatobium* and *Trichuris trichiura*, with *Ascaris lumbricoides* eggs being the predominant contaminant (60 per cent of lettuce, 55 per cent of cabbage and 65 per cent of spring onions showed *Ascaris lumbricoides* eggs).

Biologically, the highest health risk from pathogens is infections due to helminth contamination. Because helminths persist for longer periods than pathogens in the environment, host immunity is usually low to non-existent and the infective dose is small (Gaspard et al, 1997). Such microbial and parasitic contamination probably contributes to the high number of food-borne and water-related diseases in Accra such as diarrhoea (sometimes caused by typhoid or cholera), as well as intestinal worm infections. However, these also have to be seen in the context of generally sub-optimal sanitary conditions in parts of the metropolis (Arde-Acquah, 2002).

Pesticide residues on lettuce leaves

Table 6.2 shows pesticide detection prevalence and residues recorded on lettuce leaves, with maximum residue limits (MRL) as comparators. Only

Table 6.2 *Pesticide residue detection and concentrations on lettuce (n = 60)*

Pesticide	Lettuce with detected pesticide residues (%)	Range of concentrations (mg/kg) on lettuce with residues	Mean value (mg/kg) lettuce	MRL[1] (mg/kg) lettuce
Lindane	31	0.03–0.9	0.3	0.01
Endosulphan	36	0.04–1.3	0.4	0.05/0.5
Lambda-cyhalothrin	11	0.01–1.4	0.5	1.0/0.1
Chlorpyrifos	78	0.4–6.0	1.6	0.05/0.5
DDT	33	0.02–0.9	0.4	0.05

[1]MRL, maximum residue limit (Pesticide Safety Directorate, 2005).

14 per cent of samples had no detectable pesticide residues. More than 60 per cent of the lettuce samples had two or more pesticide residues, with 78 per cent of samples showing chlorpyrifos, an organophosphate of moderately acute hazard (WHO, 2005). Chlorpyrifos was the only pesticide with higher levels in one city, Kumasi. In most cases, pesticide residue levels observed exceeded the MRL.

The results of tests for pesticide residue indicate that several pesticides (particularly chlorpyrifos) are widely used by vegetable producers in Ghana, in keeping with other studies (Okorley and Kwarteng, 2002). As also described by Danso et al (2002), farmers mix cocktails of various pesticides to increase their potency. Vegetables are often eaten raw so it is not surprising to read about evidence of chlorpyrifos contamination such as can be found in waakye, a popular Ghanaian dish (Johnson, 2002). Lindane and endosulphan are restricted for the control of capsids on cacao trees, stem borers in maize and for pests on coffee plants, while DDT is banned in Ghana. However, the data show clearly that these potent agrochemicals are used irrespective of whether approved for vegetable production or not. It has been reported that in several African countries the legislation on importation and regulation of pesticides is sketchy, nonexistent or imbedded in bodies of legislation indirectly related to pesticides (Tallaki, 2005). Because of the lack of proper regulations, organochloride pesticides banned in industrialized countries for their retention in the environment or their high toxicity are still commonly used.

This widespread pesticide contamination, often exceeding the MRL, indicates potential health risks to consumers. Washing vegetables before consumption is highly recommended, but the majority of pesticides cannot just be washed away and may still pose health risks (Volpi, 2005). A rough calculation helps to elucidate this potential: The acceptable daily intake (ADI)[3] of chlorpyrifos, for example, is $0.01 \, mg \, kg^{-1}$ body weight (WHO, 1997). To exceed the ADI, a child weighing 30 kg would have to consume at least 0.3 mg of chlorpyrifos per day. With a residue level of 1.6 mg

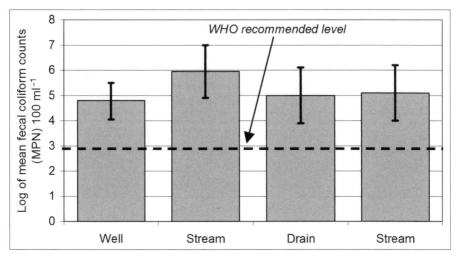

Figure 6.2 *Fecal coliform contamination levels of irrigation water used during the study period (piped water was excluded because no fecal coliforms were detected during the study period)*

chlorpyrifos kg^{-1} lettuce, the child would have to eat close to 200 g lettuce per day. The amount of lettuce (usually served with other staples, e.g. rice) is usually less than 30 g daily. However, if a child was malnourished they might be more susceptible; moreover, fetal (via maternal) or chronic neurodevelopmental effects might occur, since they are not always included in MRL analysis.

Contamination pathway study

FC levels of irrigation water used

With the exception of piped water, FC contamination levels of irrigation water from different sources used in the study significantly exceeded a geometric mean of 1×10^3 $100\,ml^{-1}$ (WHO recommended level for unrestricted irrigation[4]). For details, see Figure 6.2. Significantly higher FC counts were recorded in water samples from streams than from shallow wells in Kumasi.

The maximum FC contamination densities for shallow well and stream samples in Kumasi were 4×10^6 $100\,ml^{-1}$ and 4×10^8 $100\,ml^{-1}$, respectively, while drain and stream samples in Accra showed densities of 9×10^6 $100\,ml^{-1}$ and 2×10^7 $100\,ml^{-1}$, respectively. This corresponds well with data reported by several researchers (Cornish et al, 1999; Mensah et al, 2001; Keraita et al, 2002) in the same study area.

Helminth egg contamination levels of irrigation water used

No eggs were recorded in piped water during the study period. Arithmetic mean[5] helminth egg contamination levels in irrigation water from drains,

Table 6.3 *Mean numbers of helminth eggs in irrigation water from different sources*

City	Irrigation water source[1]	Helminth eggs (Γ^{-1}) (arithmetic mean)
Accra	Drain	8
	Stream	15
Kumasi	Shallow well	2
	Stream	27

[1] Piped water was not included because no helminth eggs were observed during the study period ($n = 15$ for each irrigation water source).

streams and wells in Accra and Kumasi also exceeded the WHO recommended level of one or less eggs per litre for unrestricted irrigation (Table 6.3). The main species of helminth eggs isolated in water and on lettuce included those of *Ascaris lumbricoides* (most predominant species in all samples), *Hymenolepis diminuta*, *Fasciola hepatica*, *T. trichura* and *Strogyloides* larvae. The results from Kumasi showed a higher helminth population in stream water than in shallow wells, probably due to run-off.

The results of the microbiological quality of irrigation water confirm earlier reports (Cornish et al, 1999; Mensah et al, 2001; Keraita et al, 2002) that low-quality water is being used for urban vegetable production in most Ghanaian cities. Shallow wells or dugouts might be expected to meet the WHO recommended standard due to the natural filtering of aquifer materials and long underground retention times. However, those used in this study were often not protected against surface inflow.

The significant differences recorded in FC levels between the two sources in Kumasi may suggest that shallow well water may pose relatively less risk to farmers and consumers, although the coliform levels still exceed 1000 counts per 100 ml. Similar results have been reported from Kenya (Hide et al, 2001). However, the situation can also change. For example, Cornish et al (1999) recorded in Kumasi temporarily higher FC population in shallow wells than in nearby streams. This may be due to the fact that probably the wells used were shallower and got more easily contaminated through surface run-off on the field (Drechsel et al, 2000).

Microbiological quality of lettuce at different entry points along the production consumption pathway

Irrespective of the irrigation water source, mean FC levels on produce exceeded recommended standards. Figure 6.3 shows FC populations on lettuce samples collected in Kumasi at the farm gate, wholesale market and retail outlets over a 12-month period and for three irrigation water sources. There were no significant differences in average lettuce contamination levels at different entry points (farm, wholesale market and retail outlet). Similar contamination levels were recorded for the three irrigation water sources in Accra (Amoah et al, 2007). High levels of FC counts (usually above the

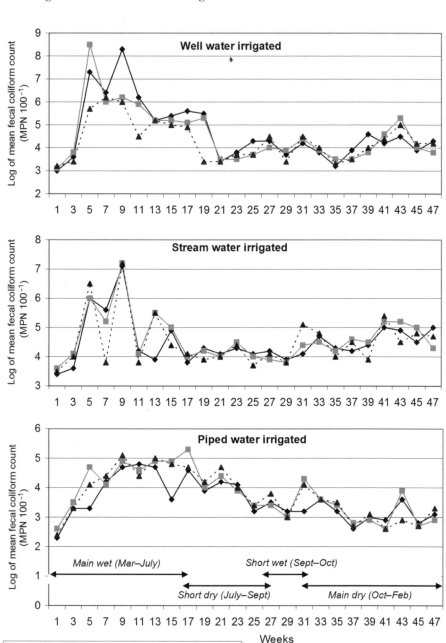

Figure 6.3 *Fecal coliform counts at different entry points on the production–consumption pathway for irrigated lettuce using water from (A) wells, (B) streams and (C) piped water in Kumasi*

common acceptable standard of 1×10^3 $100\,\mathrm{g}^{-1}$ wet weight) were recorded on all irrigated lettuce including those irrigated with piped water.

Apart from lettuce from Accra that was stream-water irrigated, higher levels of FC contamination were recorded in the rainy season than in the dry season. The difference was significant ($P < 0.05$) only in the cases of well- and stream-water irrigated vegetables from Kumasi (Amoah et al, 2005). The results also showed that in 80–90 per cent of the weeks' produce sampled in Accra and Kumasi, there was no significant difference in the FC counts of samples analysed from the farm gate to the markets and final retail points.

Helminth eggs

Helminth eggs, including those of *Ascaris lumbricoides*, *Hymenolepis diminuta*, *T. trichuris*, *F. hepatica* and *Strongyloides* larvae, were counted on lettuce samples at the different entry points. The helminth egg population ranged from one to six eggs $100\,\mathrm{g}^{-1}$ wet weight. In the majority of cases, significantly higher levels ($P < 0.05$) were recorded in lettuce that was irrigated with polluted water than those from piped water irrigated sources. However, mean helminth egg population on lettuce from the same original stock and irrigation water source did not show any significant difference from field to market (see Table 6.4).

The study revealed that the contamination of lettuce with pathogenic micro-organisms does not significantly increase through post-harvest handling and marketing, in contrast to earlier work (Armar-Klemesu et al, 1998). These results were unexpected because of the alarmingly unhygienic conditions on market sites, which are related to washing habits, display and handling of food and the availability of sanitation infrastructure. Only 31 per cent of the

Table 6.4 *Helminth egg contamination at different entry points along the production–consumption pathway*

City	Irrigation water source	Mean* helminth egg concentration (100 g^{-1} wet weight)		
		Farm	Wholesale market	Retail
Kumasi	Well	4^{a**}	5^a	4^a
	Stream	6^b	5^a	5^a
	Piped water	2^c	2^b	1^b
Accra	Drain	6^a	6^a	5^a
	Stream	4^b	3^b	4^{ab}
	Piped water	3^b	2^b	3^b

*Mean numbers represent the mean of all the different types of eggs, as well as *Strongyloides* larvae ($n = 15$ for each irrigation water source). Eggs were rounded to the nearest whole number.
**The same letters indicate no significant difference in column numbers between water sources for each city ($P > 0.05$). For differences between farm, wholesale market and retail, see Figure 6.3.

markets in Accra have a drainage system and only 26 per cent have toilet facilities, while 34 per cent are connected to pipe-borne water (Nyanteng, 1998). The results suggest that the initial on-farm contamination is so high that it overshadows any possible post-harvest contamination.

Significantly higher levels of FCs in well- and stream-water irrigated lettuce were recorded in Kumasi during the rainy season than in the dry season. A similar but non-significant trend was observed in Accra. In fact, it might be expected that in the rainy season, when contamination of the irrigation water is more diluted and farmers irrigate less, FC population on lettuce at the farm gate would be lower than in the dry season. Another observation was that crop contamination also takes place under irrigation with piped water. One reason could be that poultry manure does not dry sufficiently; another reason is that splashes from the soil during heavy rains may contain pathogens from already contaminated soil (Bastos, 1992).

Phase IV: Field trial

The results of the field trial (Amoah et al, 2005) confirmed that, even at the farm level, wastewater is only one of several sources of crop contamination. The soil and poultry manure were identified as other sources of microbiological contamination, although wastewater is the most significant one. The need to reduce the potential health risks resulting from FC and helminth contamination of urban and peri-urban vegetables thus requires a more holistic approach rather than concentrating solely on wastewater.

Health risk reduction

The study indicates it is crucial to reduce health risks associated with the consumption of contaminated lettuce at the farm level through good agricultural practices. This, however, is not as easy as is often suggested. Farmers might not be able to make changes to irrigation methods, timing and crops (Drechsel et al, 2002). Therefore, other options are under investigation under the Challenge Program on Water for Food (CPWF), a major international effort being conducted by the Consultative Group on International Agricultural Research (CGIAR). Although first trials by IWMI show that the on-farm contamination levels can be reduced through minor changes in practices, it is unlikely that contamination can be minimized below the threshold for safe consumption, as data from the use of piped water show. Thus, it will always be necessary to wash the crops, in addition to improving farming practices.

The last stage in the production–consumption chain, where food for consumption or fast-food for street sales is prepared, appears to be a good entry point for improving hygiene. This is because awareness for food safety is generally high at this point. Also, more than 90 per cent of the food vendors and consumers wash salad. However, individual methods vary widely and seldom meet the required standards (Amoah, in press). Moreover, consumers reduce health risks by trusting only food vendors with neat appearance and visually clean food (Olsen, 2006), which is a first step but not sufficient to avoid contaminated food completely (Mensah et al, 2002).

Impacts or outcomes associated with the work

The study has provided relevant information and advice to city authorities such as the Ministry of Food and Agriculture (MoFA) and the Metropolitan Assembly on wastewater use in urban and peri-urban vegetable production in Ghana. Authorities are aware of the need to have a more holistic approach in addressing issues related to the use of wastewater in peri-urban agriculture (UPA) rather than application of 'hard' solutions such as banning the use of wastewater, which are often impractical and difficult to enforce. In view of this, the authorities at the MoFA and the Metropolitan Assembly have requested assistance in the formulation of more appropriate policies on UA in the future.

Researchers have seen the need to do more research on non-treatment options for health risk reduction (e.g. CPWF and IWMI on such concepts as farm risk-reduction techniques, vegetable washing techniques). The results have contributed to the formulation of the new WHO guidelines on wastewater use in agriculture which include several non-treatment options for health risk reduction (WHO, 2006).

Leading farmers and sellers involved in the activities of the study have become aware of the risks involved in the use of wastewater and are taking

Figure 6.4 *Farmer irrigating with lowered watering can*

Figure 6.5 *Seller displaying vegetables in the market*

the necessary precautions. For example, these farmers have reduced contact with wastewater during fetching it by not stepping into the water. To further decrease contamination from the soil, watering cans are lowered during irrigation (see Figure 6.4) to reduce contamination from splash. Also, only mature compost is applied. The sellers have stopped displaying their produce on the floor, now choosing raised shelves and tables (see Figure 6.5). These practices are expected to be introduced to other farmers and sellers at the same sites as well as throughout the cities.

CONCLUSIONS AND RECOMMENDATIONS

The research demonstrates that typical microbiological and pesticide contamination levels of vegetables in Ghanaian markets pose a threat to human health. It has also shown that the majority of microbial and parasitological contamination of lettuce produced from urban sources in Accra and Kumasi comes from wastewater, application of manure and residual soil contamination. Results also show that the post-harvest sector is probably a relatively minor contributor to lettuce contamination.

Although there is enough general epidemiological evidence in support of disease transmission through the consumption of contaminated vegetables that are irrigated with wastewater (Shuval et al, 1986), it is difficult for farmers and municipalities to comply with the WHO wastewater irrigation guidelines (WHO, 2006) for various reasons (Drechsel et al, 2002). Simply banning the use of wastewater for unrestricted irrigation would deprive many farmers and sellers of their livelihood and drastically reduce the amount of many perishable vegetables in Ghana's cities (Cofie and Drechsel, 2007). It also would not solve the problem of post-harvest contamination. Therefore, more efforts are needed to test possible options for risk reduction at farm, market, street restaurant and household level (Drechsel et al, 2002). For example, market sellers could improve their vegetable washing to better meet the demands of their customers (Drechsel et al, 2000). Further requirements are that the sanitary facilities in markets provide permanent access to clean running water and that support through education and awareness campaigns takes place. Risk reduction should also include sensitization of consumers (especially parents) on potential health implications of unhygienic handling and consumption of contaminated vegetables.

In conclusion, adoption of the 'multiple barrier approach' (WHO, 2006), where health risk intervention methods (such as good irrigation practices and vegetable washing before food preparation at the household) are used, appears to be the most appropriate overall risk-reduction strategy.

ACKNOWLEDGEMENTS

This study was made possible by financial support from IDRC's International Graduate Research Awards in Urban Agriculture (AGROPOLIS), the Challenge Program on Water and Food (CPWF) projects CP38 and CP51, and a capacity building grant of the International Water Management Institute (IWMI). We are grateful to Mr Osei Tutu from the Biological Science Department of the Kwame Nkrumah University of Science and Technology, KNUST, Ghana, and Richmond Kofi Yawson, a laboratory assistant at IWMI lab, for their support in data collection and laboratory analysis.

NOTES

1 Philip Amoah, Ph.D. Biological Sciences, IWMI West Africa Office, PMB CT 112 Cantonments, Accra, Ghana; tel: +233-21-7847452 or +233-20-8154651; email: p.amoah@cgiar.org
2 Products pass through three main sampling stages from the field where they are harvested to the retail outlet where consumers buy them: the farm, where samples of crops are collected; the wholesale market, where samples are taken from crops wholesalers will purchase; and the retailer, where samples were taken two to three hours after vegetables had been displayed and where some had been refreshed.
3 The acceptable daily intake (ADI) is a measure of the quantity of a particular chemical in food which, it is believed, can be consumed on a daily basis over a lifetime without harm.

4   World Health Organization's (WHO, 2006) recommended level for uses that include crops likely to be eaten raw.
5   Arithmetic mean, which included *Strongyloides* larvae.

REFERENCES

Abbott, C. and Hasnip, N. (1997) *The Safe Use of Marginal Quality Water in Agriculture – A Guide for the Water Resource Planner*, HR Wallingford and DFID, UK, p59

American Public Health Association, American Water Works Association, Water Environment Federation (APHA, AWWA, WEF) (2001) *Standard Methods for the Examination of Water and Wastewater*, 22nd edition, Washington DC, USA

Amoah, P. (in press) 'Report on a study to test and improve the efficacy of common home-washing methods for vegetables'

Amoah, P., Drechsel, P. and Abaidoo, R. C. (2005) 'Irrigated urban vegetable production in Ghana: sources of pathogen contamination and health risk-reduction', *Irrigation and Drainage*, vol 54, pp49–61

Amoah, P., Drechsel, P., Abaidoo, R. C. and Henseler, M. (2007) 'Irrigated urban vegetable production in Ghana: Microbiological contamination in farms and markets and consumer risk group', *Journal of Water and Health*, vol 5, no 3, pp455–466

Amoah, P., Drechsel, P., Abaidoo, R. C. and Klutse, A. (in press) 'Effectiveness of common and improved sanitary washing methods in selected cities of West Africa for the reduction of coliform bacteria and helminth eggs on vegetables', *Tropical Medicine and International Health*,

Arde-Acquah, A. A. (2002) 'Sanitation and health-related problems of street-vended foods in Accra', in SADAOC *Street Food in Ghana: Types, Environment, Patronage, Laws and Regulations*, Proceedings of a roundtable conference on 6 September 2001, ISSER, University of Ghana/WILCO, Accra, pp57–60

Armar-Klemesu, M., Akpedonu, P., Egbi, G. and Maxwell, D. (1998) 'Food contamination in urban agriculture: vegetable production using wastewater', in Armar-Klemesu, M. and Maxwell, D. (eds), *Urban Agriculture in the Greater Accra Metropolitan Area*, Final Report to IDRC (project 003149), University of Ghana, Noguchi Memorial Institute, Accra, Ghana

Bastos, R. K. (1992) 'Bacteriological aspects of drip and furrow irrigation with treated wastewater', Thesis report submitted in accordance with the requirements for Ph.D, Department of Civil Engineering, University of Leeds, UK, December 1992

Clarke, E. E., Levy, L. S., Spurgeon, A. and Calvert, I. A. (1997) 'The problem associated with pesticide use by irrigation workers in Ghana', *Occupational Medicine*, vol 47, pp301–308

Cofie, O. O. and Drechsel, P. (2007) 'Water for food in the cities: The growing paradigm of irrigated (peri)-urban agriculture and its struggle in sub-Saharan Africa', *African Water Journal*, vol 1, no 2, pp23–32

Cornish, G. A., Mensah, E. and Ghesquire, P. (1999) 'An assessment of surface water quality for irrigation and its implication for human health in the peri-urban zone of Kumasi, Ghana', Report OD/TN 95, September 1999, HR Wallingford, UK

Danso, G., Drechsel, P. and Fialor, S. C. (2002) 'Perception of organic agriculture by urban vegetable farmers and consumers in Ghana', *Urban Agriculture Magazine,* vol 6, pp23–24

Drechsel, P., Abaidoo, R. C., Amoah, P. and Cofie, O. O. (2000) 'Increasing use of poultry manure in and around Kumasi, Ghana: Is farmers' race consumers' fate?', *Urban Agricultural Magazine* 2, pp25–27

Drechsel, P., Blumenthal, U. J. and Keraita, B. (2002) 'Balancing health and livelihoods: Adjusting wastewater irrigation guidelines for resource-poor countries', *Urban Agriculture Magazine*, vol 8, pp7–9

Environment Canada (2006) 'Municipal wastewater sources and characteristics', www.atl.ec. gc.ca/epb/issues/wstewtr.html (accessed June 2006)

Gaspard, P., Ambolet, Y. and Schwartzbrod, J. (1997) 'Valorisation des boues de stations d'épuration en vue de l'amélioration des sols destines a l'agriculture: Contamination parasitaire et modélisation en vue de la gestion du risqué sanitaire', *Bulletin de L' Académie National de Médicine*, vol 181, no 1, pp43–138

Ghana Statistical Services (GSS) (2002) *2000 Population and Housing Census; Summary Report of Final Results*, Accra, Ghana

Hide, J. M., Hide, C. F. and Kimani, J. (2001) 'Informal irrigation in the peri-urban zone of Nairobi-Kenya. An assessment of surface water quality for irrigation and its implication for human health in the peri-urban zone of Kumasi, Ghana', Report OD/TN 105, March 2001, HR Wallingford, DFID, UK

IMCSF (1974) Micro-organisms in foods. Sampling for microbiological analysis: Principles and specific applications. The International Commission on Microbiological Specifications for Food, University of Toronto Press, Toronto, Canada

Johnson, P. N. T. (2002) 'Overview of a recent DFID/NRI/FRI project on street-vended food in Accra', in SADAOC (2002) *Street Food in Ghana: Types, Environment, Patronage, Laws and Regulations*, Proceedings of a roundtable conference on 6 October 2001, ISSER, University of Ghana/WILCO, Accra, pp41–51

Keraita, B., Drechsel, P., Huibers, F. and Raschid-Sally, L. (2002) 'Wastewater use in informal irrigation in urban and peri-urban areas of Kumasi, Ghana', *Urban Agriculture Magazine*, vol 8, pp11

Mensah, E., Amoah, P., Drechsel, P. and Abaidoo, R. C. (2001) 'Environmental concerns of urban and peri-urban agriculture: Case studies from Accra and Kumasi', in Drechsel, P. and Kunze, D. (eds), *Waste Composting for Urban and Peri-urban Agriculture: Closing the Rural-Urban Nutrient Cycle in Sub-Saharan Africa*, IWMI, FAO, CABI, pp55–68

Mensah, P. D., Yeboah-Manu, K., Owusu-Darko, K. and Ablordey, A. (2002) 'Street foods in Accra, Ghana: How safe are they?' *WHO Bulletin*, vol 80, pp546–554

Ntow, W. J. (2001) 'Organochlorine pesticides in water, sediment, crops, and human fluids in a farming community in Ghana', *Archives of Environmental Contamination and Toxicology*, vol 40, pp557–563

Nyanteng, V. K. (1998) 'Draft summary report on food markets and marketing in the Accra metropolis', in *Report on the National Seminar on Food Supply and Distribution Systems*, AMA/FAO, Accra, 1998

Okorley, E. L. and Kwarteng, J. A. (2002) 'Current status of the use of pesticides in urban and peri-urban vegetable production in the central region of Ghana', Paper presented at RESEAU Ghaneen/SADOAC Workshop on Sustainable Food Security in West Africa, 30–31 January 2002, Milkin Hotel, Accra

Pesticide Safety Directorate (2005) Excel Spreadsheet of current MRLs. Department for Environment, Food, and Rural Affairs, London, UK. Available at: www.pesticides.gov.uk/uploadedfiles/web-Assets/PSD/MRL-Spreadsheet.xls (accessed August 2005)

Schwartzbrod, J. (1998) *Methods of Analysis of Helminth Eggs and Cysts in Wastewater, Sludge, Soils and Crops*, Université Henry Poincare, Nancy, France

Shuval, H. I., Fattal, B. and Yekutiel, P. (1986) 'State of the art review: an epidemiological approach to the health effects of wastewater reuse', *Water Science and Technology*, vol 18, pp147–162

Tallaki, K. (2005) 'The Pest-control System in the Market Gardens of Lomé, Togo', in Mougeot, L. J. A., *Agropolis: The Social, Political and Environmental Dimensions of Urban Agriculture*, Earthscan Publications, London, UK

Tchobanoglous, G., Franklin, B. and Stensel, D. (1991) *Wastewater Engineering: Treatment, Disposal and Reuse*, Metcalf and Eddy Inc., New York

Volpi, A. M. (2005) 'Pesticides, can we avoid them?', www.annamariavolpi.com/pollutants. html (accessed June 2006)

WHO (2005) *The WHO Recommended Classification of Pesticides by Hazard and Guidelines to Classification: 2004*, WHO, Geneva, Switzerland

WHO (2006) *Guidelines for the Safe Use of Wastewater, Excreta and Grey Water: Wastewater Use in Agriculture*, vol 2, WHO, Geneva, Switzerland

7

Water Contamination and its Impact on Vegetable Production in the Rímac River, Peru

Henry Juarez[1]

INTRODUCTION

The rapid growth of population in Lima is leading to an expansion of unplanned informal settlements which lack many basic urban services such as waste management, clean water and drainage. Combined with poor farming practices, these settlements create the perfect conditions for large amounts of run-off. This run-off contains waste products, for instance from overflowing septic systems, that affect surface waters, the environment and ultimately the health of food producers and consumers. The latter face risks from waterborne diseases, while the farmers are also affected by skin ailments and intestinal problems. Due to the scarcity of clean water and lack of adequate treatment of domestic sewage, the use of contaminated water is a common practice which sustains farmers' livelihoods in urban and peri-urban areas. Lima is merely one of many international cities located at the mouth of a large river, many of which face the same environmental problems that directly impact on water quality. Typically, such waters are polluted by excessive quantities of nutrients, plus they are contaminated with pathogens and toxic chemical substances that affect both the ecosystem and the public's health (Lee-Smith and Prain, 2006).

The Rímac River basin is one of the most important in Peru. It provides enough drinking water for approximately 60 per cent of Lima's population of 7.7 million people. The river supplies a large population with a wide range of socio-economic activities including mining, an industry established long ago in the upper and middle part of the basin. In addition, hydroelectric generation occurs along the river, as well as agricultural irrigation.

Water contamination of the Rímac River historically has been related to the discharge of mining waste in the upper and middle part of the basin

(Castro, 1993; Infante and Sosa, 1994; MEM-DGAA, 1997; Bedregal et al, 2002). In addition, the disorderly growth of human settlements around areas of crop production – where 35 per cent of the vegetables marketed in Lima are produced – aggravates this situation, contaminating the horticultural products with enteric pathogens (Castro and Sáenz, 1990; Moscoso and León, 1994; Moscoso, 1998; Acosta et al, 2001; Manrique et al, 2002a, b).

To date, it has not been clear whether vegetable production sites irrigated with contaminated water have also been affected by heavy metals (HM) and enteric pathogens to the same magnitude. Information about plant uptake of HM, patterns of water quality in the river basin and causes and sources of contaminants are lacking, as well as farmers' awareness of the risks inherently involved in their work. Furthermore, there are significant public health risks associated with food consumption where urban agriculture (UA) and peri-urban agriculture (UPA) use contaminated water. Given the lack of information on related health risks, this study was initiated to assess and provide guidance to stakeholders at different levels, from producers to consumers.

Research objectives

This research involved an analysis of HM and fecal contamination in the Rímac River basin to determine the environmental risks and the impact on soil, water and vegetables in the eastern part of Lima. The objectives of the present study were to first study historical data (spatial and temporal) on the quality of water in the basin. The specific objectives were to document and analyse the sources of pollution in the river basin both currently and in the past; to map the spatial distribution of HM and fecal contamination in the entire Rímac River basin using secondary data; and to determine whether municipal and national environmental regulations have influenced a reduction in the pollution.

The second major objective was to evaluate existing environmental risks affecting agricultural land, water and vegetables produced. Specifically, the study aimed to understand farmers' perceptions concerning water quality used in vegetable irrigations; to characterize the actual levels of HM and fecal contamination in water located both in the main weir and irrigation canals; to determine the environmental risks due to the absorption of HM in the soil and the risks to human health due to concentrations of HM and fecal contamination in vegetables; and to propose recommendations to improve the quality of agriculture products.

METHODS

Assessment of historical water quality of the Rímac River
Sources of information
HM such as arsenic (As), cadmium (Cd), chromium (Cr) and lead (Pb) were selected because they are known to be contaminants that can be absorbed by

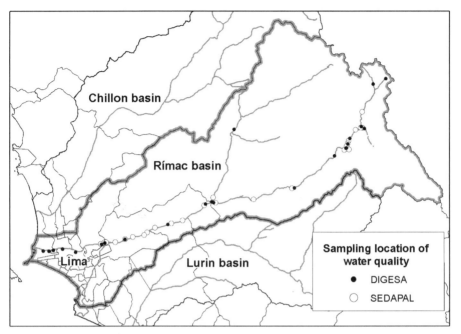

Figure 7.1 *Sampling location of water quality in the Rímac River basin conducted by SEDAPAL (white dots) and DIGESA (black dots)*

plants, enter the food chain and steadily accumulate in the organs of human beings, resulting in chronic poisoning (Zhou et al, 2000). The World Health Organization (WHO) guidelines on wastewater identify fecal coliform (FC) bacteria as an indicator of fecal pollution in wastewater used in agriculture and aquiculture (Mara and Cairncross, 1989).

The study included monthly data collection from the General Environmental Health Directorate of the Ministry of Health (DIGESA) on Cd, Cr, As, Pb and FC from 24 sampling stations between 2000 and 2004. Also, we used annual data from the Drinking Water and Sewerage Service of Lima (SEDAPAL) on Cd, As and Pb from 28 sampling stations between 1996 and 2004. It was difficult to obtain monthly data from SEDAPAL due to political issues surrounding As contamination in water. Figure 7.1 shows the sampling stations from both sources used in this research.

Mapping spatial–temporal water quality of the Rímac River

Each sampling station was geo-referenced through a global positioning system (GPS). Spatial and temporal surfaces of water quality were created using spatial analysis from ArcGIS 9.0.[2] Annual average numbers for each sample station were used for interpolation. A buffer of 2000 m around the Rímac River was defined in order to delimit the interpolated values, and the inverse distance weighted (IDW) tool was used to interpolate the parameters for water quality.

Table 7.1 *Water quality index based on Class III stipulated by the Peruvian law for vegetable irrigation and animal consumption*

Parameter	Water quality index			
	Low	Medium	High[a]	Very high[a]
As (mg/L)	<0.01	0.01–0.02	0.02–0.05	>0.05
Cd (mg/L)	<0.01	0.01–0.05	0.05–0.10	>0.10
Cr hexane (mg/L)	<0.05	0.05–1.00	1.00–2.00	>2.00
Pb (mg/L)	<0.01	0.01–0.05	0.05–0.10	>0.10
Fecal coliforms (MPN/100 ml)	<100	100–1000	1000–10,000	>10,000

[a] Above permissible limits for vegetable irrigation. MPN, most probable number.

To assess the contamination status of water, the General Water Law of Peru was consulted (MINAG, 1969, 1976). A water quality index was designed to provide a simple and concise method for expressing the water quality of the river. The index was classified in four levels, taking into account the stipulations for water used for vegetable irrigation and animal consumption (known as Class III waters) as determined by the Peruvian General Water Law. Low and medium indexes consider values below the maximum permissible limits set for Class III, while a high index considers values two to ten times higher than the maximum permissible set for Class III. A very high index considers values that are extremely high which are set for Class III, but which are unsuitable for vegetable irrigation (Table 7.1).

Water quality maps were prepared of each of the water quality parameters sampled. While they are based on average annual values, and hence do not give us detailed information about monthly variation, they offer a general idea of what is influencing their concentrations across the Rímac River.

Perception of risk of pollution or exposure to contaminants

A detailed survey of 125 of the 276 livestock- and vegetable-producing households had been previously carried out in the Lurigancho Chosica district.[3] Survey interviewers asked questions on characteristics of members of the household, the context of the livestock activity, agronomic management, livestock management, post-harvest and market activities, water use, excreta and solid waste disposal, complementary activities, expenditures and family income, division of labour, and institutional and organizational networks that are assisted by the activities of UA (Lozano, 2004). We used this survey to understand what farmers think about their exposure to pollution or contaminants.

Assessment of water, soil and crop quality in eastern Lima

Lurigancho Chosica is one of the most important farming areas, supplying as much as 35 per cent of the vegetable market in Lima. It is located in the lower

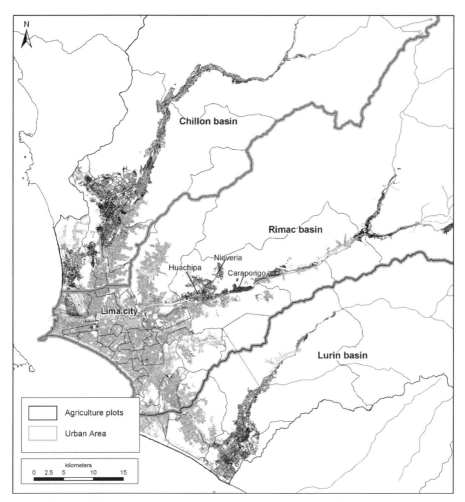

Figure 7.2 *Urban and peri-urban agriculture in and around Lima (agricultural plots shown in black, urban areas in grey) and three research locations: Carapongo, Nievería and Huachipa*

watershed of the Rímac River in the eastern part of the city. Three of its four vegetable production sites were selected in this study to assess the impact of irrigation water on soils and in vegetables. The three sites selected for study were Carapongo, Nievería and Huachipa (Figure 7.2). Site selection was chosen because of the extent of the farms (871 of 1115 ha of all vegetable production areas are located in these three sites). In addition, land use change is occurring very rapidly in the fourth farming area (Ñaña) because of the conversion of agricultural land into permanently built-up areas which are covered with infrastructure (Juarez et al, 2007).

Water quality of irrigation canals

Twenty-five water samples were collected in both October 2004 and March 2005 in Carapongo for HM and FC analysis. Additionally, 20 water samples were collected in Huachipa and Nievería in September 2005 for FC. Sampling sites included main and lateral irrigation canals, water gates and proximity to populated places. Criteria included samples taken during both the dry and the wet seasons.

Water samples for HM were preserved with nitric acid for subsequent laboratory analysis. Water samples for FC were collected in sterilized glassware containers. Each sample was geo-referenced by a GPS location. To assess the contamination status of water, the General Water Law of Peru was consulted (MINAG, 1969, 1976) and the indexes displayed in Table 7.1 were used to classify all maps.

Quality of vegetables

Vegetables represent the main crop around Carapongo, Nievería and Huachipa. In total, 57 vegetable samples were collected and sampled.

Carapongo was divided in five agricultural zones to allocate samples proportionally (Campo Sol, Guadalupe, Huancayo, Nuevo Horizonte and Tulipanes). Five vegetable samples within each zone were collected in December 2004 (FC analysis) and April 2005 (FC and HM analysis). Crops included were those noted in previous surveys conducted in the area (Lozano, 2004): huacatay (*Tagetes minuta* L., used as seasoning or as a condiment in Peruvian foods), lettuce (*Lactuca sativa* L.), radish (*Raphanus sativus* L.), turnip (*Brassica rapa* L. var. *rapa*) and the common beet (*Beta vulgaris* L. var. *crassa*). An additional seven samples were collected in Nievería and Huachipa in March 2006 for FC analysis. Selected crops were turnip, common beet, huacatay, celery (*Apium graveolens* L.) and grass (*Lolium perenne* L.).

Five to six sub-samples were taken for each vegetable sample during harvest season. For the assessment of HM, vegetables were classified in accordance with their edible parts: roots (turnip, radish and common beet) and foliage (lettuce, huacatay and celery). In addition, vegetables were collected before and after washing for FC assessment.

The freshly harvested vegetables were brought to the laboratory and washed, initially with running water to remove the soil particles, followed by three washings with distilled water. Samples were cut into small pieces before being oven-dried at 105°C to constant weight. The samples were then pulverized with a mortar and subjected to wet digestion in the conical flask with HNO_3 and $HClO_4$ (2:1) to extract total HM (AOAC, 1984). Then 10 ml of HCl was added to dissolve inorganic salts and oxides. Care was taken to prevent contamination in all steps of the process. HM in all samples were determined with atomic absorption spectroscopy (AAS) for Cd and Cr, AAS + graphite furnace for Pb, and AAS + hydride generation for As.

Guidelines for HM in vegetables were taken from several sources including the 1995 Chinese National Standards Analytical Methods (Zhou et al, 2000) and Codex Alimentarius (2006). The guidelines for HM were set at 0.50 mg As/kg, 0.20 mg Cd/kg, 0.50 mg Cr/kg and 0.3 mg Pb/kg in fresh matter. FC bacteria count was analysed by the five-tube most probable number method (MPN) (APHA, 1992). Limits for FC were taken from those recommended by the WHO (1989).

Heavy metals in soils

Soil samples were obtained from the same place as vegetable samples. All soils were sampled manually with a soil sampler (Model J, Spectrum Technologies).[4] Samples were extracted from topsoil (approximately 20 cm depth) and air-dried for four days. Tiny roots and other residues were removed before the soil was ground and sieved in a 2-mm mesh. Fractions less than 2 mm were analysed for total as well as dissolved HM.

Total HM content was determined using an acid mixture. One gram of soil for each sample, in duplicates, was transferred into a 200-ml digestion flask. Ten millilitres of a mixture of concentrated $HClO_4$ and HNO_3 (ratio of 2:1) was added before covering the digestion flask with a watch glass. The mixture was heated progressively and boiled under reflux for two hours after which the digestion flask was cooled (Baker and Amacher, 1982; AOAC, 1984). Tests for available HM in soils were carried out with 1N ammonium acetate shaking for two hours (Bradi, 1984; Yanai et al, 1998). Heavy metals in all samples were determined with AAS for Cd and Cr, AAS + graphite furnace for Pb and AAS + hydride generation for As.

Guidelines for HM concentrations in soils were taken from the Taiwan standards for assessment of soils contaminated with HM (Wang et al, 1994; Chen et al, 1996). These included A values (the upper limit of background concentration), B values (the acceptable level) and C values (the intervention level, at which pollution control is needed). These guidelines take into account

Table 7.2 *The Taiwanese standards for assessment of soils contaminated with heavy metals*

Element	A value		B value		C value	
	0.1 M HCl extracted (ppm, dry soils)	*Total concentration (ppm, dry soils)*	*0.1 M HCl extracted (ppm, dry soils)*	*Total concentration (ppm, dry soils)*	*0.1 M HCl extracted (ppm, dry soils)*	*Total concentration (ppm, dry soils)*
As	–	16.0	–	30.0	–	40.0
Cd	0.4	2.0	11.0	4.0	2.0	5.0
Cr	12.0	100.0	25.0	250.0	40.0	400.0
Pb	18.0	50.0	150.0	300.0	200.0	500.0

A value, reference top value of the background range; B value, further monitoring level; C value, pollution control level.
Source: Chen, 1992; Chen et al, 1996.

total HM in soils and available HM for plants (extracted by 0.1 M HCl) (Wang et al, 1994). The C values were used as guidelines for assessment of HM in soils.

Statistical analysis of the data

Data were subjected to an analysis of variance and means were compared by Fisher's protected least significance difference (LSD) test. Statistical analysis was carried out with assistance of CIPSTAT7 (CIP Statistical Analyser) developed at the International Potato Center. Descriptive statistics were used to explain the basic features of the data in a study to provide simple summaries about the samples.

RESEARCH FINDINGS AND DISCUSSIONS

Historical contamination of Rímac River

The incorporation of spatial–temporal data on water quality in a geographical information system (GIS) proved to be a useful tool in assessing pollution trends for different metals in different parts of the watershed and the associated potential risks. As the maps track changes over time, they are useful in identifying trends and offer a general idea of what is influencing their concentrations throughout the Rímac River. Looking at these summary maps is faster and easier than doing extensive detailed analysis of the raw water-quality data. Also, they help to focus the analysis directly on areas of concern, locate areas of high risk due to the presence of HM and FC and assesses the discrepancies between different data sets.

Figure 7.3 shows the annual averages over time of As, Cd, Cr, Pb and FC reported by SEDAPAL (1997–2004) and DIGESA (2000–2004), while Figure 7.4 shows the spatial distribution of As, Pb and FC along the Rímac river corresponding to the year of greatest contamination (2000 or 2001) and the last reported (2004).

Cd and Cr did not affect any part of the basin and were always below the maximum permissible level for vegetable irrigation. However, vegetable growing areas in the lower part of the basin were affected with contaminated water for at least two years with As (2000 and 2002) and for the entire evaluated period with Pb (1997–2004).

As expected, the presence of Pb and As in the Rímac River is related to discharges of mining wastes in the upper and middle part of the basin (Infante and Sosa, 1994; MEM-DGAA, 1997). The data also indicates that As levels have reduced in the last two years (2003–2004) (Figure 7.5A). This is probably due to an obligatory implementation of the Environment Adequacy Program (Programa de Adecuación Ambiental, PAMA), as applied to ongoing mining operations and the Environment Impact Assessments which are applied to new mine operations (MINEM, 1993). However, there is no evidence that water quality has been improved in terms of Pb (Figure 7.5B). High levels of

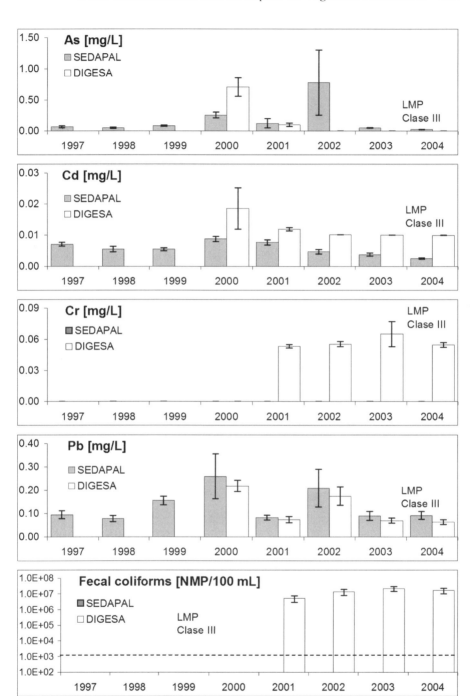

Source: SEDAPAL (1997–2004) and DIGESA (2000–2004)

Figure 7.3 *Annual means for As, Cd, Cr, Pb and FC in the Rímac River, where the vertical lines represent the standard error of the mean*

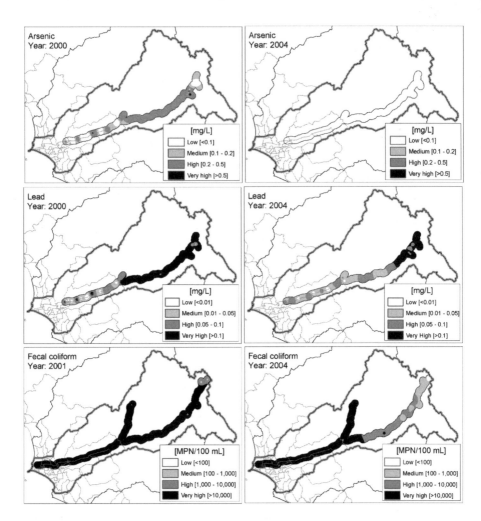

Figure 7.4 *Spatial patterns of water contamination with As, Pb and FC in the Rímac River for the year of greatest contamination (2000 or 2001) and the most recent year (2004)*

Pb still exist in the Rímac River basin which could pose an important risk for agricultural activity in the Rímac Valley. Figure 7.6 shows the degree of mining activity established long ago in the basin.

A different situation regarding microbiological contamination was observed. FC were found to be higher than $10^5/100$ ml in almost the entire river except for a small section of the highest part of the basin (see Figures 7.3 and 7.4). These findings were also reported in other studies (Castro and Sáenz, 1990; Moscoso and León, 1994; Moscoso, 1998; Acosta et al, 2001; Manrique et al, 2002a, b). These high levels of FC present in the river are due to the

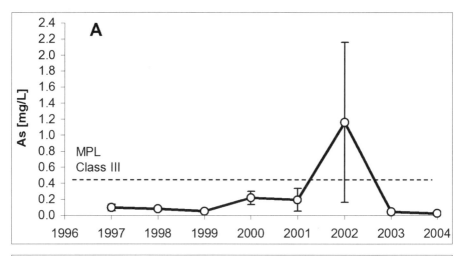

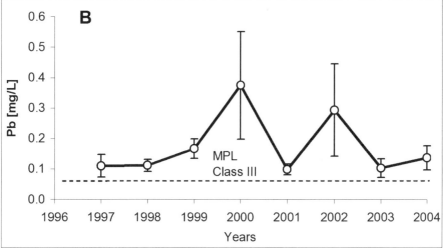

Figure 7.5 *As and Pb in the upper part of the basin from 1997 to 2004, where vertical lines represent the standard error of the mean*

inadequate treatment of domestic sewage from human settlements along the Rímac River. Improper disposal thus contaminates the surface water. As an indicator bacteria, the presence of such a level of FC contamination suggests that the water poses a risk for agricultural activity in the Rímac Valley.

Perception of risk to pollution or exposure to contaminants

Among survey participants, 74 per cent of farmers perceived that water used for irrigation is contaminated and 73 per cent thought domestic sewage and

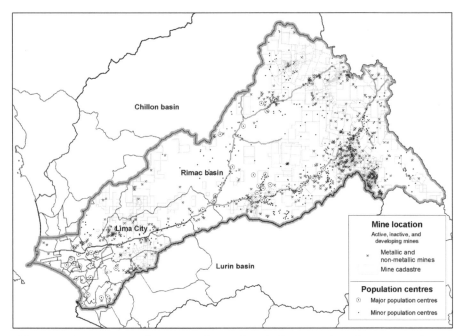

Figure 7.6 *Location of mines and populated places in the Rímac River basin*

solid waste from residential communities comprised the principal sources. Eleven per cent of farmers thought that waste from mining activities affected water quality and were aware that HM can affect soils, crops and human health. The appraisal of how farmers perceive the risk to pollution or exposure of contaminants can be very complex (Grasmück and Scholz, 2003). These results show that farmers' perception of water quality is attributed mainly to factors that are easily observed such as domestic sewage (excreta) and solid waste from urban communities (such as bottles and papers). If the exposure is not visible (e.g. the presence of parasites and pathogens or HM in irrigation water or the accumulation of HM in soil or plants), the perception of the risk is less apparent.

Assessment of water, soil and crop quality in eastern Lima

Water quality of irrigation canals

None of the water samples collected from the canals exceeded the maximum permitted limits for As, Cd, Cr and Pb according to Codex Alimentarius (2006). Thus the water is suitable for use in irrigating vegetables from the standpoint of HM.

Contamination of irrigation water with pathogens and parasites is a more serious problem for vegetable production. The Rímac River is a main source of contamination due to the sewage and excreta effluents that are emptied directly into irrigation canals that supplement farms. More than 97 per cent of water samples taken from irrigation canals were found to contain five million FC, which is above the maximum permitted limits. The quality of irrigation water worsened downstream as the river passed through more densely populated settlements in the Rímac Valley.

Heavy metal contents in soils

From the analysis of the soils it was observed that more than 40 per cent of the samples contained high levels of As and Cd, as well as Pb that exceeded the safe limits according to Chen (1992), Chen et al (1996) and Wang et al (1994). The total concentration (Figure 7.7) indicates the location of potential contamination pathways, such as soil ingestion by children, inhalation of dust, soil adhesion on edible leaves and other sources associated with handling the soil (Nabulo, 2002). Extractable concentrations of HM in soils are considered to be indicators of availability to plant roots (Kimberly and William, 1999). Observations in this study (Figure 7.8) showed that uptake by plants was only a fraction of the total HM concentrations (Sauve et al, 2000). Because there are no guidelines for safe levels of available As in soils, it was impossible to compare amounts.

Some of the available Cd measured in soil solution may have come from the weathering of parent rock, as it was also observed that 68 per cent of the variability of available Cd came from total Cd in soils ($P < 0.001$). The levels of trace elements present in soils result from the interactions between the geology of the parent rock and the soil-forming factors and human activities (Kabata-Pendias and Adriano, 1995). However, available As found in the soil may have originated from pesticides, fertilizers or sewage sludge (Alloway and Ayres, 1993).

The results also suggest different amounts of HM in soils across evaluation sites. Localities near the main gate of Nuevo Horizonte accumulated more Cd ($P < 0.001$) and Pb ($P = 0.019$) than other sites that were far from the gate.

Quality of vegetables

In spite of the mechanisms involved in elemental uptake by roots (non-metabolic or metabolic), plants are known to respond to the amounts of available inorganic nutrients in soil solution (Li and Shumas, 1996; Baban, 1999; Madrid et al, 2002). Analysis of some of the vegetables in the irrigated fields showed that both Cd and As were accumulated more in foliage than in roots. But only the indigenous aromatic plant huacatay exceeded the maximum permitted levels (guideline: 0.50 mg As/kg and 0.20 mg Cd/kg) (Figure 7.9). The plant is used in small quantities for seasoning or as a condiment in various Peruvian foods and so may not pose a health risk to consumers.

Seventeen per cent of lettuce and 31 per cent of radish samples had very high levels of FC contamination, presumably due to exposure to the highly contaminated canal water (WHO, 1989). Contamination loads in the two

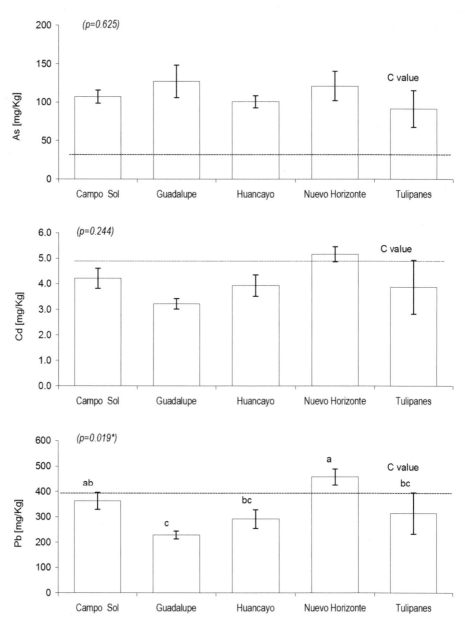

P indicates statistical differences between sites (*significant, **highly significant) and lower-case caption indicates differences across sites.
Source: Contamination value in soils (C values) were taken from Chen (1992), Wang et al (1994) and Chen et al (1996).

Figure 7.7 *Total heavy metals in soils. Vertical lines represent the standard error of the mean*

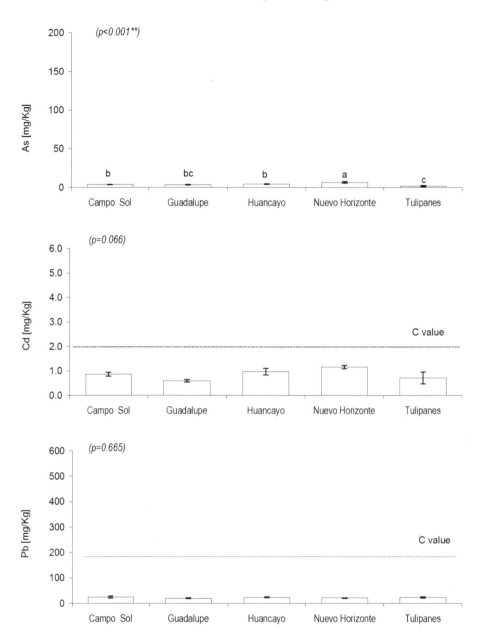

P indicates statistical differences between sites (*significant, **highly significant) and lower-case caption indicates differences across sites.
Source: Contamination value in soils (C values) were taken from Chen (1992), Wang et al (1994) and Chen et al (1996).

Figure 7.8 *Available heavy metals in soils. Vertical lines represent the standard error of the mean*

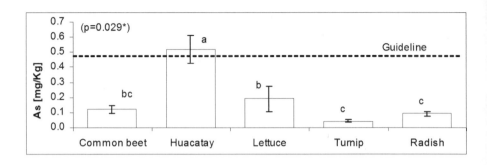

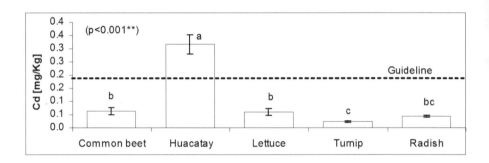

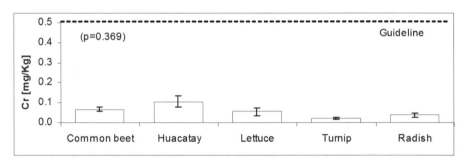

P indicates statistical differences between crops (*significant, **highly significant) and lower-case caption indicates differences across crops.

Figure 7.9 *Total heavy metals in vegetables. Vertical lines represent the standard error of the mean*

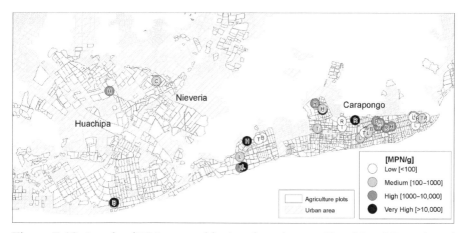

Figure 7.10 *Levels of FC in vegetables in selected areas: Huachipa, Nievería and Carapongo*

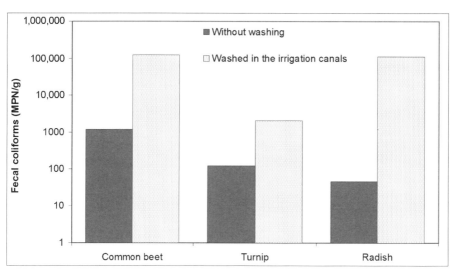

Figure 7.11 *Effect of washing vegetables in the irrigation canals*

vegetables varied depending on the location of sampling (Figure 7.10). Washing vegetables with irrigation water from the canal increased the level of contamination, and 57 per cent of clean vegetables were found to be contaminated after the washing process (Figure 7.11).

CONCLUSIONS AND RECOMMENDATIONS

This research exemplifies the use of multiple data sources and multiple pathway analyses to capture different ways through which food gets contaminated. The integration of spatial–temporal data on water quality in a GIS model proved to be a useful tool to evaluate pollution trends for different contaminants flowing through a watershed and downstream into a coastal city, and the associated risks of contamination of vegetable-growing areas. The project demonstrated the usefulness of participatory mapping (GIS) as an effective approach to dissemination of environmental pollution information and risk reduction strategies. These methods also will help the development of appropriate information dissemination packages for the communities who depend for their livelihood upon these ecosystems.

Vegetables grown in areas contaminated by water from the Rímac River had high levels of As and Pb, although a reduction in As has been observed over the last two years due to improvements in the Ministry of Energy and Mines' regulatory systems. Regulations now exist which involve obligatory programmes of environmental adequacy as well as environmental impact assessments (MINEM, 1993). Despite the high levels of Pb found in the river water, the content found in soils and crops does not pose a significant health risk. However, significant levels of As and Cd were found in soil, and these could be harmful. Investigation of the Cd and As levels in vegetables showed that there was more accumulation in foliage than in roots. The indigenous aromatic plant huacatay contains Cd and As concentrations that are above those permitted by Peruvian law. Contamination of irrigation water with pathogens was a more serious problem for vegetable production. FC were found to reach levels higher than permitted by Peruvian law: more than 97 per cent of water samples from irrigation canals contained five million MPN/ 100 ml, which is beyond the maximum permitted limits for FC bacteria. Analysis of lettuce and radish samples showed 17 per cent and 31 per cent of the samples having bacterial loads that are above permitted limits. The practice of washing vegetables using water from the irrigation canals further contaminated these vegetables: 57 per cent of clean vegetables were contaminated during the rinsing process.

To complement the chemical and biological data collection, the study sought to understand how farmers perceive environmental risks. Not surprisingly, farmers tended to depend on observed phenomenon rather than hearsay references to mining risks. Almost three-quarters associated domestic sewage and other urban solid wastes in the irrigation channels, with only 11 per cent mentioning mining. Provision of correct information about the sources of pollution in the environment would go a long way to enhancing awareness of such risks to the community.

The high levels of microbiological contamination raises the need for measures to be applied which would make it possible to improve the quality of water used for irrigation and washing in the study area. While the ideal scenario would be to prevent untreated domestic sewage from entering the river through sewerage, this is not realistic at the moment. Consequently, other mechanisms that permit in situ quality improvement should be considered, such as the use of washtubs of clean water for vegetables or the implementation of reservoirs so as to treat irrigation water.

Based on this research, efforts were made to find an appropriate technology for a simple, low-cost way to improve water quality. Several small reservoirs have been built to test their feasibility for reducing potential contaminants in irrigation water. The cleaning process is quite simple and greatly assisted by solar radiation and increased temperature. First, water enters the reservoir, where it is retained for about 10–14 days. While there, the water is exposed to physical processes that act on the pathogens and parasites. For instance, the action of keeping the suspended bacteria isolated from their biological cycle for more than ten days reduces their concentration and viability. At the same time, parasites sink to the bottom of the reservoir and eventually die, leaving the water clean for vegetable irrigation.

Furthermore, the use of small reservoirs has been shown to have additional benefits. Introducing productive fish to the reservoir compensates for the loss of agricultural land and provides families with additional nutritious food to consume or sell. In addition, this aspect makes the concept of reservoirs appealing for other farmers in the area who might wish to construct one on their land.

ACKNOWLEDGEMENTS

My sincere thanks go to the International Potato Center (CIP), the Urban Harvest Initiative and in particular to the International Development Research Centre (IDRC), whose AGROPOLIS grant programme funded the field research. I would specially like to thank my research supervisor Gordon Prain from Urban Harvest and Julio Moscoso from the Centro Panamericano de Ingeniería Sanitaria y Ciencias del Ambiente (CEPIS) for their time and effort in helping me develop a research product. They provided excellent scientific guidance pertaining to their specialties. I also want to thank my thesis advisor, Victor Aguilar from Universidad Nacional Agraria la Molina, for his advice. Finally, I want to thank Reinhard Simon for providing the freedom and opportunity to pursue this research as part of my activities at CIP.

NOTES

1   Henry Juarez, MSc, Environmental Sciences, PO Box 1558, Lima 12, Peru, tel: +51-1-3175311 ext 2156; email: h.juarcz@cgiar.org
2   Copyright 2002 ESRI. 380 New York Street, Redlands, CA 92373-8100, USA.
3   Urban Harvest baseline survey (Huachipa, Ñaña, Nievería y Carapongo)
4   www.specmeters.com/Soil_Moisture/Soil_Samplers.html
5   http://riuweb.cip.cgiar.org/cipstat/

REFERENCES

Acosta, L., Alvaro, H., Jiménez, H., Manrique, R. and Torralba, R. (2001) *Proyecto Regional Sistemas Integrados de Tratamiento y Uso de Aguas Residuales en América Latina: Realidad y Potencial*, Estudio General del Caso, Villa el Salvador, Lima, Peru

Alloway, B. J. and Ayres, D. C. (1993) *Chemical Principles of Environmental Pollution*, Oxford University Press, Oxford, UK

APHA (1992) *Standard Methods for the Examination of Water and Wastewater*, 18th edition, American Public Health Association, Washington, DC, USA

Association of Official Chemists (AOAC) (1984) *Official Methods of Analysis*, 14th edition, Sidney Williams (ed), AOAC, Arlington, VA, USA

Baban, S. M. (1999) 'Use of remote sensing and geographical information systems in developing lake management strategies', *Hydrobiologia*, vol 395–396, pp211–226

Baker, D. E. and Amacher, M. C. (1982) 'Nickel, copper, zinc and cadmium', in Page, A. L., Miller, R. H. and Keeney, D. R. (eds) *Methods of Soil Analysis, Chemical and Microbiological Properties*, SSSA Publishers, NY, USA, pp323–336

Bedregal, P., Montoya, E., Torres, B., Olivera, P., Mendoza, P., Ubillus, M., Ramos, R., Creed-Kanashiro, H., Penny, M., Aleman, M., Gil, A., Junco, J. and Ganoza, L. (2002) 'Determination of trace elements and heavy metals in agricultural products cultivated at the River Rimac in the city of Lima', Co-ordinated research project on use of nuclear and related analytical techniques in studying human health impacts of toxic elements consumed through foodstuffs contaminated by industrial activities, Report on the First Research Co-ordination Meeting, Vienna, Austria, 18–22 March 2002, International Atomic Energy Agency

Bradi, M. A. (1984) 'A study of comparative behaviour of selected heavy metals in polluted and unpolluted estuarine and coastal sediments', Ph.D. Thesis, University of Lancaster, Lancaster, UK

Castro, M. L. (1993) *Evaluación de Riesgos para la Salud por el uso de las Aguas Residuales en Agricultura: Aspectos Toxicologicos*, Centro Panamericano de Ingeniería Sanitaria y Ciencias del Ambiente (CEPIS)

Castro, M. L. and Sáenz, R. (1990) *Evaluación de Riesgos para la Salud por el uso de Aguas Residuales en Agricultura; Aspectos Microbiológicos*, Centro Panamericano de Ingeniería Sanitaria y Ciencias del Ambiente (CEPIS)

Chen, Z. S. (1992) 'Metal contamination of flooded soils, rice plants, and surface waters in Asia', in Adriano D. C. (ed), *Biogeochemistry of Trace Metals*, Lewis Publishers, FL, USA, pp85–107

Chen, Z. S., Lee, D. Y., Lin, C. F., Lo, S. L. and Wang, Y. P. (1996) *Contamination of Rural And Urban Soils in Taiwan*, Boston, London

Codex Alimentarius (2006) 'Codex General Standard for Contaminants and Toxins in Foods', CODEX STAN 1993–1995, Rev.2–2006, www.codexalimentarius.net/download/standards/17/CXS_193e.pdf (accessed December 2007)

Grasmück, D. and Scholz, R. W. (2003) 'Risk perception of heavy metal soil contamination by high-exposed and low-exposed inhabitants', ETH–UNS Working Paper No 42, Zurich, Switzerland

Infante, L. and Sosa, S. (1994) 'Cuantificación espectrofotométrica de arsénico en aguas de consumo humano en la vertiente del Río Rímac', tesis presentada a la Universidad Nacional Mayor de San Marcos, Facultad de Farmacia y Bioquímica para el Grado de Químico Farmacéutico

Juarez, H., Prain, G. and Castro, M. (2007) 'The use of spatial analysis in urban and peri-agriculture research and development in Lima', Working Paper, International Potato Centre, Lima, Peru

Kabata-Pendias, A. and Adriano, D. C. (1995) 'Trace metals', in Rechcigl, J. E. E. (ed) *Soil Amendments and Environmental Quality*, Lewis Publishers, Boca Raton, FL, USA, pp139–168

Kimberly, M. F. and William, H. (1999) 'Trace metals in Montreal urban soils and the leaves of *Teraxacum officinale*', *Canadian Journal of Soil Science*, vol 79, pp385–387

Lee-Smith, D. and Prain, G. (2006) 'Urban Agriculture and Health', in Hawkes, C. and Ruel, M. T. (eds) *Understanding the Links Between Agriculture and Health. International Food Policy Research Institute (IFPRI)*, Washington, DC, USA

Li, S. and Shumas, L. M. (1996) 'Heavy metals movement in metal-contaminated soil profiles', *Soil Science*, vol 161, pp656–66

Lozano, M. (2004) *Documentacion de las bases de datos de la encuesta de linea base del Cono Este: Nieveria, Carapongo, Huachipa, Ñaña*, Cosecha Urbana – Centro Internacional de la Papa (CU-CIP)

Madrid, L., Diaz-Barrientos, E. and Madrid, F. (2002) Distribution of heavy metal content of urban soils in parks of Seville, Chemosphere vol 49, pp1301–1308

Manrique, R., Jiménez, H., Álvaro, H., Torralba, R. and Acosta, L. (2002a) 'Estudio de Viabilidad: Sistema de Aprovechamiento de las Aguas Residuales en el Fundo San Agustín, Callao – Perú', Proyecto Regional Sistemas Integrados de Tratamiento y Uso de Aguas Residuales en América Latina: Realidad y Potencial, Convenio IDRC–OPS/HEP/CEPIS 2000–2002, vol 86

Manrique, R., Jiménez, H., Álvaro, H., Torralba, R. and Acosta, L. (2002b) 'Estudio complementario del caso fundo de San Agustín, Callao – Perú', Proyecto Regional Sistemas Integrados de Tratamiento y Uso de Aguas Residuales en América Latina: Realidad y Potencial, Convenio IDRC–OPS/HEP/CEPIS 2000–2002

Mara, D. and Cairncross, S. (1989) *Guidelines for the Safe Use of Wastewater and Excreta in Agriculture and Aquaculture*, Organización Mundial de la Salud, Geneva, Switzerland

MINAG (1969) 'Ley general de aguas', MINAG–Dirección General de Aguas, Decreto Ley 17752

MINAG (1976) 'Ley general de aguas y sus reglamentos', MINAG–Dirección General de Aguas

MINEM (1993) 'Reglamento para la protección ambiental en las actividades minero metalúrgicas', Decreto Supremo No 016–93-EM, Diario Oficial el Peruano

Ministerio de Energia y Minas (Dirección General de Asuntos ambientales) (MEM-DGAA) (1997) 'Evaluación ambiental territorial de la cuenca del Río Rímac', MEM-DGAA, www.minem.gob.pe/archivos/dgaam/publicaciones/evats/rimac/rimac.htm (accessed December 2007)

Moscoso, J. (1998) Acuicultura con aguas residuales tratadas en las lagunas de estabilización de San Juan, Lima, Perú, Centro Panamericano de Ingeniería Sanitaria y Ciencias del Ambiente (CEPIS), Lima, Peru

Moscoso, J. and León, G. (1994) *Uso de Aguas Residuales*, Hojas de Divulgación Técnica, Centro Panamericano de Ingeniería Sanitaria y Ciencias del Ambiente (CEPIS), Lima, Peru

Nabulo, G. (2004) 'Assessment of heavy metal uptake by selected food crops and vegetables around Kampala city area, Kampala, Uganda', Department of Botany technical report submitted to IDRC–Agropolis, Makerere University, Kampala, Uganda

Sauve, S., Henderson, W. and Allen, H. E. (2000) 'Solid-solution partitioning of metals in contaminated soils: Dependence on pH, total metal burden, and organic matter', *Environmental Science and Technology*, vol 34, pp1125–1131

Wang, Y. P., Chen, Z. S., Liu, W. C., Wu, T. H., Chaou, C. C., Li, G. C. and Wang, T. T. (1994) 'Criteria of soil quality – establishment of heavy metal contents in different categories', EPA/ROC, Grant No EPA-83-E3H1–09–02

WHO (1989) 'Health guidelines for the use of wastewater in agriculture and aquaculture', Report of the World Health Organization Scientific Group, WHO Technical Report Series: No 778, Geneva, Switzerland

Yanai, J., Yabutani, M., Yumei, K., Huang, B., Luo, G. and Kosaki, T. (1998) 'Heavy metal pollution of agricultural soils and sediments in Liaoning Province, China', *Soil Science and Plant Nutrition*, vol 44, pp367–375

Zhou, Z. Y., Fan, Y. P. and Wang, M. J. (2000) 'Heavy metal contamination in vegetables and their control in China', *Archives of Environmental Contamination and Toxicology*, vol 16, no 2, p239

8

Wastewater Use and Urban Agriculture in Kinshasa, DR Congo

Kifuani Kia Mayeko[1]

INTRODUCTION

Urban agriculture is an important source of food and jobs for people in cities throughout the world, especially in developing countries. In addition, urban agriculture helps keep a city clean by using urban waste to improve and fertilize land in and around the city for growing crops. In 2006, the UN announced that the world had reached the benchmark where more than 50 per cent of people are now living in cities. This justifies a look at the role UA can play in providing food for the demographic explosion. The contribution of UA to food production at the world level is growing; however, despite this importance, few governments take measures to protect it. Moreover, it is often forgotten during the formulation of national policy.

The Democratic Republic of Congo covers an area of 2,345,000 km². There are approximately 52 million residents of which 49 per cent are men and 51 per cent women (Ministère du Plan, 2004). Human development indicators show that severe poverty exists not only in rural zones, but also in towns and cities. Only 45 per cent of the country's population has access to safe drinking water (UNPD, 2001; UNICEF, 2002) and nearly 80 per cent of the population lives on less than US$0.20 per person per day (Ministère du Plan, 2004). In the face of this extreme poverty, UA helps contribute to the food security of populations in Kinshasa and in other cities of the country, as it does in other developing countries (Delvaque, 1980; Agripromo, 1985; Niang, 1996; Mokili, 1998).

Description of study area

The Kinshasa region includes the Malebo Pool, a lake-like expansion of the Congo River, with its channel islands, vast alluvial plain that extends from Maluku to Kintambo and the Bateke Plateaux (Fluoriot et al, 1975; Van Caillie,

1983). The N'sele and N'djili Rivers are both important tributaries of the Congo River in the region of Kinshasa (Fluoriot et al, 1975; Van Caillie, 1983). Kinshasa is surrounded by numerous local rivers which drain the hills encircling the city and influence its climate. Kinshasa has a humid tropical climate characterized by a rainy season and a dry season (Crabbe, 1980; Van Caillie, 1983).

Kinshasa occupies $9965.2\,km^2$, of which only $600\,km^2$ is residential land, indicating that the city has a vast peri-urban area where urban development has yet to occur (de Saint-Moulin, 1976; INS, 1991). The city is divided into 24 communes and has a population of five million (Ministère du Plan, 2004), including more than 40,000 market gardeners.

The formal large-scale practice of market gardening in Kinshasa began in 1954–1956, when the Belgian agronomist Voldeker established the first two market-gardening centres in Kinshasa – the N'djili and Kimbanseke centres – to produce vegetables and exotic fruits destined for Europeans living in Kinshasa (Symposium des Organizations et Dynamiques Maraîchères, 2004). In 2004, the Ministry of Rural Development reported that 90 per cent of the vegetables consumed in Kinshasa are produced by market gardeners living in Kinshasa (Symposium des Organizations et Dynamiques Maraîchères, 2004).

Concepts, Objectives and Hypothesis

Mougeot (2005) has defined UA as an industry that is localized inside a city (intra-urban) or on the periphery (peri-urban) and distributes a variety of food and non-food products, uses human and material resources and provides products and services to satisfy the needs of an urban population. According to Autissier (1994), market gardening is the production of annual plants (shrubby or herbaceous) in a delimited agrarian space, generally exploited intensively, whose yield is sold in fairly large quantities and may be used as ingredients for sauces or salads. Several studies have described the organization of UA in various countries (Due, 1986; Lele, 1986; Barry, 2005; Tallaki, 2005). Hovorka (2005) and others have described the importance of analysing gender when studying UA.

The water needed for urban farming is often provided by both raw and treated wastewater in many developing countries (Mara and Cairncross, 1991). While it is ideal that water used for farming be treated before it is used, the effectiveness and scale of treatment often limits the availability of safe irrigation water. Farmers are then exposed to health risks as they end up using lower-quality waters. Wastewater can be treated in numerous ways, and the type of treatment determines the final use of the wastewater (Eckenfelder, 1982; Bontoux, 1983; Mukana and Kifuani, 2000; Kifuani, 2004).

This study had the following objectives:

1   Identify the structure, organization, characteristics and constraints of UA in Kinshasa

2 Define interventions responding to the needs of market gardeners in the *Centre Agricole de Kinshasa* and define the specific roles, responsibilities and actions of the actors involved in resolving conflicts linked to UA
3 Determine the bacteriological and chemical characteristics of irrigation water that is used and its seasonal variations
4 Evaluate the use of certified compost of Projet de Compostière de Quartier (PCQ) in producing vegetables (notably *Amaranthus hybridus*)
5 Propose alternative solutions for sustainable UA.

The following hypotheses were proposed:

- Market gardeners of Kinshasa, like the rest of the Congolese population, are poor; they are exposed to chemical and microbial contamination because of the poor quality of irrigation water and their agricultural practices; and they are limited in terms of organization, training and assistance.
- UA is gender specific (in agricultural practices, social and economic levels); it is also seasonal.
- The water used for irrigation does not meet the acceptable bacteriological and chemical levels for irrigation or drinking water and exhibits seasonal variations.
- Composting with certified PCQ compost of the Red Cross – Congo improves the yield of vegetables produced.

METHODS

Research on UA in Kinshasa was undertaken using the multi-actor ecosystem participation approach (MEPA). According to this approach, UA is thought of as an ecosystem, with different actors and physical milieus involved in complex and multiple interactions. The actors are market gardeners, civil society, the private sector, public power and researchers. The market gardener is the central actor of the agro-ecosystem. Civil society represents organized non-governmental organizations (NGOs) or individuals and households who influence the agri-ecosystem by sustaining agricultural production. The private sector comprises all micro enterprises or enterprises influencing UA through inputs and outputs. Public power is represented by the state as the regulator of land in the city, sometimes to the benefit and other times to the detriment of UA. Researchers look for solutions to the problems of UA through observation, experiments or scientific analyses. Physical milieus are places where UA activities such as production, transportation, marketing and consumption take place. Interactions between the elements of this agri-ecosystem are material and observable or virtual and non-observable. Furthermore, these interactions are categorized as 'synergistic' if they promote the safe practice of UA or 'antagonistic' if they limit it.

A team of 13 people (five women and eight men) was assembled; the team included five market gardeners, four of whom were heads of groups of market gardeners on the site. The remaining people represented the other actors of the agri-ecosystem. Before the survey, the team received a short training session in various aspects of the project such as the questionnaire, concepts of agriculture, investigative techniques and the psychological aspects of the investigation. An initial site visit was undertaken to see how the Centre Agricole de N'djili was structured and to identify the actors involved in UA.

The investigation of the Centre Agricole de N'djili market gardeners was conducted with a probability survey in sampling and a focus group. Each interviewee was asked identical questions and received the same explanations, allowing the information to be collected in a consistent manner in the farmer survey.

Questionnaire for market gardeners

A questionnaire was designed containing 109 questions regarding the structure of market gardeners and the organization of UA. Before the final questions were selected, a field investigation was conducted on 14 market gardeners, chosen at random, to collect relevant information and to test their semantic comprehension of typical questions. Their replies were used to improve the final questionnaire, which was then reviewed by psychologist academic advisors. The questionnaire was written in French but administered in the local language, Lingala.

Sampling area, sample size and parameters

The survey site was determined after collaboration with the management committee of the Centre Agricole de N'djili. The sampling area consisted of 594 market gardening plots. The plot was chosen as the unit of sampling and analysis for the survey, and each plot was assigned an identification number. Market gardeners in the sample area are distributed in four groupings: blocks A, B, C and D. The sample size (N) was fixed at 303 plots corresponding to 303 market gardeners. The parameters identified for inclusion in the questionnaire are given opposite.

Testing irrigation water

Choice of sampling sites and sample collections

Ten sampling sites (see Table 8.1) were chosen on the basis of geographical location.

Samples of water for chemical analyses were collected according to the protocol described in the literature (Rodier, 1984). For each site, one sample was collected, usually during the second week of the month, for 12 months.

Samples for bacteriological analyses were collected from the seven most frequented sites (ENDA, EPP, EET, ERO, EPA, EMA and EPNA). Water samples were collected in borosilicate glass bottles that had been washed and rinsed with distilled water and then sterilized in an autoclave at 120°C to a pressure of $150\,kg/cm^2$ for 30 minutes. In most cases, samples were collected

SURVEY PARAMETERS

Social – name, gender, age, religion, civil status, family composition, place of residence, level of education

Economic – household's standard of living, daily expenses, possession of essential goods ensuring the quality of life (electricity, water, radio, TV, refrigerator, freezer)

Technical and production – sources of irrigation water, organization of market gardening (seniority, area exploited, time devoted, distribution of tasks by gender) and agricultural equipment and inputs

Biodiversity – types of vegetable species cultivated, animal species raised

Environmental (other than biodiversity) – bacteriological and chemical characteristics of irrigation water, management of agricultural inputs affecting the environment, risks from behaviour during work, health risks to market gardeners or other actors in the agri-ecosystem

Self-reliance – identification of problems related to UA, research on causes of problems and their solutions, assistance for the market gardener, identification and analysis of solutions by the market gardener, definition of important needs by the market gardener, strengthening of capacities.

each Monday, from 08:30 to 10:30 am, for six consecutive weeks. After collection, samples were placed in an ice-cooled tank and then transported to the laboratory and kept in a refrigerator.

Field analyses

Preliminary tests for pH, electrical conductivity, total dissolved solids and temperature were conducted in the field after each sample collection. The pH

Table 8.1 *Water sampling sites in greater Kinshasa*

Sampling site	Abbreviation	Water types	Abbreviation
N'djili River: downstream	ENDA	Enhanced well	EPA
N'djili River: upstream	ENDM	Crude well	EPNA
Nkelo: drainage area	ED	Tap water	ERO
Nkelo: small bridge	EPP		
Nkelo: pond	EET		
Nkelo: market	EMA		
Nkelo: experimental garden	EJE		

Location of water sample collection: ENDA, from N'djili River downstream; ENDM, from N'djili River upstream; ED, from drainage area; EPP, at small bridge; EET, from pond site; EMA, from market; EJE, from experimental garden; EPA, from built wells; EPNA, from natural wells; ERO, tap water.

was measured with a portable pH meter (pHScan 3), and the electrical conductivity, the total dissolved solids and the temperature were measured with a Sension conductivity-meter.

Laboratory analyses

The samples of irrigation water were tested for chemical and biological components, including major elements (N, P, K) from fertilizer, minor and trace elements and heavy metals. Chemical parameters were determined by classic chemical analysis methods. The total alkalis and the chemical oxygen demand were determined volumetrically; other parameters were determined colorimetrically (Berl, 1951; Hach Chemical Company, 1979; Alexeev, 1980). Lead (Pb) and cadmium (Cd) were analysed after extraction with a chloroformic solution of dithizone. Colorimetric analyses were conducted with a Hach spectrophotometer, either a model 2000 or a model 1105.

The irrigation water samples were also tested for total coliform, fecal coliform, fecal *streptococci*, *Escherichia coli* and helminths eggs. Presumptions and count tests were undertaken using the most probable number (MPN) method, determined by culturing a certain number of samples (generally five) and (or) diluting these samples. The estimation is based on the principle of dilution until extinction (Ayres and Mara, 1997; Tanawa and Djeuda, 1998; Kiyombo, 2005). *Escherichia coli* was identified with Kovac's reagent (Kiyombo, 2005), and helminth eggs were identified with the use of a binocular microscope (Olympus CH20) after centrifuging (Kubota centrifuge).

Evaluation of Projet de Compostière de Quartier compost

Compost is produced by the aerobic breakdown of vegetable wastes and provides a strong fertilizer input for crops. The need for fertilizer in the Centre Agricole de N'djili precipitated the application of compost. As part of this study, an experimental garden was planted to evaluate the effect of compost on the productivity of *Amaranthus hybridus*, one of the most prolific crops in Kinshasa (Mokili, 1998; Boole, 2000).

Six vegetable beds were fertilized with compost or with sludge (SLU) and watered with wastewater (WW) or with faucet water (FW). The vegetables were measured from the base to the last bud with a tape measure. The basal diameter was measured with vernier callipers. A tape measure was also used to measure the length of leaves from the stem to the tip. Vegetables were divided into three sample zones, each with an area of $1\,m^2$, and then weighed.

RESEARCH FINDINGS

Typology and location of urban agriculture in Kinshasa

In Kinshasa, urban farmers can be found in mountain basins, residential plots, along main roads and in the hills, as well as in peri-urban areas.

- UA in mountain basins – The geography of Kinshasa includes many streams in small gorges. UA is common in these basins. The most important sites are the Funa, Kalamu, Lukunga, N'djili and Nsanga valleys. This type of agriculture is the main focus of this study.
- UA on residential plots – UA is practised on residential plots when enough space is available. Most produce from these plots is destined for the consumption by the producer and helps to secure food for poor families. In some cases, produce is sold in the market. Constraints on this type of agriculture include the limited amount of land available for crops, the destruction of crops by poultry or other domestic animals and a lack of water for plots situated in peripheral zones, such as Kimbanseke, Kinsenso, Maluku, Mont Ngafula, N'djili and N'sele.
- UA on public thoroughfares – UA is practised along the main streets, such as the Boulevard Lumumba. In addition, various crops can be grown in small neighbourhood streets. This type of agriculture is exposed to air pollutants primarily from automobiles that can accumulate in plants.
- UA in the hills – UA is also practised in the hills of Kinshasa. This type of UA is rare and is generally undertaken by very poor families who have no other space for crops.
- Peri-urban agriculture (UPA) on the Bateke Plateaux and Plain of N'sele – Numerous farms situated on the Bateke Plateaux, the Plain of N'djili-Brassérie and at Manenga are quite large. In some cases, modern equipment and qualified personnel are used. Ploughing is often done with a tractor. Some farms have agronomist engineers. Cash crops are mostly grown on these sites, notably peanuts, cassava and corn.

Main stakeholders

There are several important relevant stakeholders related to UA in Kinshasa. First, the Ministry of Rural Development that, besides its traditional role of sustaining rural development, has a general secretariat and various services available to support UA in particular. Among the different services this ministry provides, the National Support Service for Urban and Peri-Urban Horticulture (SENAHUP) specifically concentrates on urban and peri-urban horticulture.

SENAHUP is a national support service for the development of urban and peri-urban horticulture. Their mandate is to promote urban horticulture by coordinating farming and fruit-growing activities, and provide extension services and business development services to associations of farmers. The Union of Market Garden Cooperatives of Kinshasa (UCOOPMAKIN) was established on 27 November 1987 and acts as the umbrella organization for agricultural cooperatives in the city. Each agricultural cooperative manages the interests of its members and also an agricultural centre, and each centre occupies a given market-garden perimeter and manages all the market gardeners working on-site. UCOOPMAKIN currently has 32 cooperatives, corresponding to 32 agricultural centres (Symposium des Organizations et Dynamiques Maraîchères, 2004).

Characteristics of urban agriculture in the Centre Agricole de N'djili

Social characteristics

The survey found that 55.8 per cent of farmers were younger than 50 years old, suggesting that UA is a developing activity; 77.5 per cent were married, widowed or divorced and responsible for a household; and the majority (82.5 per cent) reside in the commune of N'djili, with the difference commuting to their plots daily. The survey showed that 68.3 per cent of the market gardeners had ten years or more of experience in farming, which indicates knowledge of the trade, with some (13 per cent) having more than 40 years of professional experience. Despite this, agricultural activities were still done manually and there were very few advanced technologies employed. The hoe was found to be the most used tool.

Results show that the surveyed market gardeners were dissatisfied with the amount of assistance they received from the state or from NGOs. In almost 90 per cent of the cases, market garden activities were financed entirely by the market gardeners themselves and, as a result, the market gardeners felt a great sense of proprietorship over their farming operations. That said, many stated that they would like to have training to improve their capacity and skills.

The study found that most market gardeners sold their crop to generate income to satisfy various needs. Approximately 93 per cent sold their crops at market, while 15 per cent reserved part of their crop for family consumption and less than 1 per cent gave produce away.

Conflicts were found to exist between various market gardeners (19.4 per cent) and between market gardeners and clients (18.8 per cent). Conflicts between farmers were most often related to plot boundaries, access to water resources, perceived 'unfair' competition and problems linked to agricultural equipment or inputs. Conflicts between market gardeners and clients were caused by price, non-payment and various other misunderstandings.

A problem analysis was conducted to help market gardeners identify their problems and possible solutions. The main problems identified were: difficulties of obtaining seeds or phytosanitary products, challenges associated with procuring irrigation water, theft of vegetables during the night, lack of agricultural tools and financial support, floods, poor maintenance of access roads, infectious diseases, lack of electricity, lack of supply depots and insufficient access to manure and other fertilizers.

Accordingly, the farmers proposed a number of solutions, including: the provision of financial support, credit and incentives from the state and NGOs; the creation of jobs by the government; the organization of collective works for irrigation (laying out irrigation canals); rehabilitation of access roads; installation of dikes to keep the N'djili River from flooding their plots; installation of on-site drinking water and electricity; site security; new wells; opening of savings and loan banks; organization of a committee for the agricultural centre; treatment of irrigation water; supplying the agricultural centre with agricultural inputs; and harmonized prices. It is notable that the farmers were very focused on state-driven solutions.

In terms of education, the survey found that 34 per cent of the women were uneducated, 17 per cent possessed a primary education, 36 per cent had some secondary schooling and only 1.6 per cent had some post-secondary education.[2] In contrast, 6 per cent of the men were uneducated, 21 per cent had a primary education, 56 per cent had some secondary schooling and 11 per cent had some post-secondary education. Only 21 per cent of the group of respondents finished secondary school. The low level of education among these farmers suggests that interventions in the field should include basic education. Also, the survey revealed the majority of farmers were Christian (79 per cent).

Economic parameters

Even though most market gardeners (77 per cent) owned their own residential garden plots, most still live in very poor conditions. In fact, 60 per cent lived on residential plots shared by at least two households, where 19 per cent of the plots were shared by 4–14 households. The majority of the households (62 per cent) included more than five people, and a few families (2 per cent) were comprised of 14 members or more. Such overpopulation is a characteristic sign of poverty, creates conflicts and demands more money to satisfy the basic needs of each person in the household. Sanitary installations at most residences were traditional water closets (38 per cent) or squat toilets (35 per cent). The traditional water closet is a hole dug in the ground covered with boards or sheet metal. These installations offer no comfort, are difficult to maintain and facilitate microbial contamination. The majority of the households surveyed had drinking water (73 per cent) and electricity (86 per cent). However, 43 per cent of the households did not possess an electrical burner; instead, charcoal was used for cooking. The production of charcoal from wood contributes to the advanced deforestation observed in the Kinshasa region and produces large quantities of carbon dioxide, a greenhouse gas, not to mention a high amount of indoor pollution.

The daily household expenditures for men and women are shown in Table 8.2. Daily expenditures per person were less than US$1 for almost 90

Table 8.2 *Daily expenditure per person, 2005*

US$	Men	Women	Total	Frequency (%)
No response	2	1	3	0.99
0.00–0.50	97	68	165	54.46
>0.50–1.00	58	41	99	32.67
>1.00–1.50	12	7	19	6.27
>1.50–2.00	2	6	8	2.64
>2.00–2.50	4	1	5	1.65
>2.50–3.00	1	–	1	0.33
>3.00	2	1	3	0.99

per cent of the men and women, and such results are not gender-specific. More than 50 per cent of the men and women live in poverty on less than US$0.50 per day. Dongmo et al (2005) also note the poverty level of market gardeners in other areas.

Technical and production parameters

*   *Nutrients and phytosanitary products.* Urea (92.4 per cent) and NPK (88.5 per cent) were the most used fertilizers in the market garden in the Centre Agricole de N'djili. These were used on almost every crop and, in most cases, in excessive quantities which pollute the soil, groundwater and run-off. Nitrates and phosphates are very mobile and can reach the water table. The run-off from irrigation water or rain can contaminate surface water (Fall and Fall, 2001). The other fertilizers used were manure (72.6 per cent), brewers' grains (46.5 per cent), animal dung (42.2 per cent), sawdust (29.7 per cent), coffee hulls (29.7 per cent), compost (29.0 per cent), domestic waste (26.7 per cent) and other organic waste (8.6 per cent). The most used phytosanitary products were Thiodan (91.1 per cent), DDT (78.9 per cent) and Sumithion (18.8 per cent). The other chemical fertilizers used were aldrin (3.6 per cent), chlordane (1.3 per cent), lindane (0.9 per cent) and HCB (0.3 per cent). Less-polluting solutions such as *Titonia, Lantana camara*, tobacco powder and wood ash were also used as insecticides.
*   *Water supply and types of irrigation.* Most fields were irrigated by sprinkling with a watering can, usually of 10 l capacity. Irrigation water was taken from springs, wells or the N'djili River. Irrigation ditches were more common in blocks B and C. Block A experienced water shortages in the dry season and, in some fields, even during the rainy season. Block D was flooded during the rainy season, which destroyed the crops. Several market gardeners used well water for drinking or kitchen use.

Biodiversity parameters

*   In terms of the types of crops grown, there was little difference found between the rainy season and dry season. The differences between seasons are highlighted in Table 8.3 and result from the fact that in the rainy season some crops do not grow as well as in the dry season. In addition to vegetable crops, each field also had at least one fruit tree and 75 per cent of the fields had a palm tree. Crops were affected by some diseases, such as tracheomycosis, damping off, mildew and leaf perforation.
*   Only about 10 per cent of the market gardeners raised poultry or livestock. Animal species included pigs, chickens, ducks and goats.

Environmental parameters

pH, electrical conductivity, temperature, salinity and total dissolved solids in the irrigation water. Table 8.4 shows the average physical characteristics of irrigation water used in the Centre Agricole de N'djili plots for both the dry and rainy seasons. Each value is the average of eight samples

Table 8.3 *Vegetables grown during the dry and rainy seasons, 2005*

Species	Dry season frequency (%)	Rainy season frequency (%)
Amaranth (*Amaranthus hybridus*)	93	71
Sorrel	82	74
Hypomea sp.	78	80
Pointe noire	68	12
Lettuce	10	3
Celery	11	8
Murelle	30	8
Okra (*Abelmoschus esculentus*)	18	10
Spinach	38	17
Eggplant (*Solanum melonzena*)	15	2
Tomatoes (*Lycopersium esculentum*)	10	3
Cassava (*Manihot esculentum*)	28	26

Other crops that were only occasionally planted (less than 10 per cent in either season) include carrots, sweet potatoes, cabbage, peppers, chives and onions.

Table 8.4 *Water quality parameters during the rainy (dry) season, 2005*

Sampling site	pH	T (°C)	EC (μS/m)	TDS (ppm)	SS (mg/l)	Turb. (mg/l)	TA (mg/l)	COD (mg/l)
ENDA	6.5	26.2	13	6	77	79	57	352
	(7.3)	(24.6)	(9)	(4)	(38)	(43)	(55)	(401)
ENDM	6.7	26.9	11	6	66	73	65	320
	(7.1)	(25.6)	(11)	(5)	(38)	(41)	(56)	(444)
EPP	4.38	28.2	84	41	41	26	66	303
	(4.8)	(27.9)	(32)	(16)	(31)	(24)	(58)	(411)
EET	5.3	29.4	43	21	47	29	55	294
	(5.7)	(28.6)	(31)	(15)	(26)	(13)	(54)	(432)
EMA	6.8	29.1	46	26	43	26	47	369
	(7.2)	(27.0)	(40)	(20)	(37)	(34)	(56)	(430)
EPA	6.39	28.8	39	18	35	16	29	358
	(6.2)	(27.5)	(32)	(16)	(29)	(17)	(32)	(388)
EPNA	6.4	27.8	43	22	39	26	61	396
	(6.6)	(25.1)	(33)	(16)	(42)	(25)	(55)	(352)
ERO	6.5	28.6	12	8	33	12	36	296
	(7.3)	(26.3)	(16)	(8)	(31)	(14)	(37)	(330)
EJE	5.4	28.1	40	13	43	43	51	434
ED	5.0	26.5	47	41	31	10	50	363
	(5.6)	(26.1)	(25)	(15)	(35)	(12)	(45)	(337)
FAO standards	6.5–8.4	NA	<70,000	<450	NA	NA	NA	NA

COD, chemical oxygen demand; EC, electrical conductivity; SS, suspended solids; T, temperature; TA, total alkalis; TDS, total dissolved solids; Turb., turbidity; NA, not available.

taken during the rainy season and four samples for the dry season (corresponding to 12 months of the year). Water destined for use in the market garden should have a pH of 6.5–8.4 (FAO, 2003). The chloride concentration should be approximately 600 mg/l, and the maximum electrical conductivity should be 2500 μS/m (Niang, 1996). The results in Table 8.4 indicate that most of the water sampled met these criteria. Monthly variations in pH and temperature proved to be slight. Irrigation water was, in most cases, acidic and at optimum levels for aquatic species (Nisbet and Verneaux, 1970; Arrignon, 1991). Average values for total dissolved solids indicate that the water is not very saline.

Turbidity and suspended solids of the irrigation water. Turbidity measures the clarity of water. Suspended solids are any particles larger than 2 μm. The turbidity determined for the two seasons studied ranged from 10 to 79 mg/l for all water samples (Table 8.4). The results for turbidity and suspended solids indicated the irrigation water was polluted and, although pollution will not affect the porosity of the soil, it affects aquatic life. For example, in water where there are high levels of turbidity, fish exhibit thickened skin cells in the gills, reduced productivity and changes in behaviour (Nisbet and Verneaux, 1970; Rodier, 1984).

Chemical oxygen demand of the irrigation water. Chemical oxygen demand (COD) is an expression of the water's organic load and testing involves determining the total amount of organic matter in the water. The irrigation water of the Centre Agricole de N'djili was found to be rich in organic matter: values ranged from 294 to 444 mg/l (Table 8.4). The highest concentrations at most sampling locations were registered in the dry season, suggesting a change in farming activities or concentration affected by dehydration.

Total alkalis and sodium contents of the irrigation water. Total alkalis content is related to the presence of bicarbonates and especially to sulphates, phosphates and silicates (Rodier, 1984). The alkalinity varied noticeably according to the month, but seasonal variations were not remarkable (Table 8.4). The research found, however, that sodium (Na) concentration for all samples was very high (Table 8.5). High concentrations of sodium in the irrigation water are likely to increase the alkalinity of the soil through leaching and reduce its permeability, particularly at the surface, where crusting will be observed (FAO, 2003). The literature recommends adding calcium, such as gypsum or calcium carbonate, to amend sodium-rich soil (FAO, 2003).

Major fertilizer elements in the irrigation water. Wastewater carries nutrients that fertilize the soil; however, in excess, macronutrients can damage the environment. The average concentrations of $N-NH_4$, $N-NO_2$, $N-NO_3$, PO_4^{3-} and K^+ we found in our study are depicted in Table 8.5. These results indicated high concentrations of nitrate and potassium, low concentrations of phosphate and even lower concentrations of nitrite. The presence of NH_4^+ in water indicates organic pollution by micro-organisms, especially of fecal origin. Exposure to high doses of nitrates is accompanied by an increase in the

Table 8.5 *Mean concentrations of major and minor fertilizer elements in the irrigation water during the rainy (wet) season, 2005*

Sampling site	Concentration (mg/l)									
	NH_4^+	NO_2^-	NO_3^-	PO_4^{3-}	K^+	Ca^{2+}	Mg^{2+}	SO_4^{2-}	Cl_2	Na^+
ENDA	0.87	0.13	232	13	597	2.0	0.5	18	0.40	203
	(0.77)	(0.09)	(182)	(10)	(149)	(1.7)	(0.5)	(11)	(1.06)	(97)
ENDM	0.97	0.13	265	10	68	2.3	0.5	17	0.41	182
	(0.86)	(0.08)	(362)	(10)	(184)	(2.1)	(0.9)	(11)	(0.28)	(94)
EPP	0.42	0.08	209	19	831	1.7	0.6	16	0.30	80
	(0.45)	(0.09)	(356)	(11)	(1158)	(1.7)	(2.3)	(17)	(0.21)	(34)
EET	0.49	0.12	236	12	1237	2.8	0.8	13	0.16	37
	(0.53)	(0.22)	(280)	(15)	(754)	(2.7)	(2.2)	(11)	(0.15)	(38)
EMA	1.04	0.33	285	9	1076	1.5	0.8	14	0.26	85
	(1.11)	(0.65)	(1445)	(15)	(1433)	(2.6)	(0.9)	(17)	(0.34)	(60)
EPNA	1.28	0.27	264	13	862	2.1	0.7	14	0.21	56
	(0.45)	(0.15)	(172)	(13)	(861)	(2.8)	(1.0)	(6)	(0.22)	(46)
EPA	0.46	0.15	219	14	910	1.9	0.7	10	0.22	53
	(1.39)	(0.18)	(371)	(11)	(1326)	(2.8)	(0.76)	(17)	(0.20)	(50)
ERO	0.49	0.09	236	11	51	1.9	0.6	15	0.13	40
	(0.83)	(0.10)	(233)	(12)	(133)	(1.9)	(0.6)	(11)	(0.14)	(28)
EJE	0.29	0.12	230	10	19	0.7	0.3	18	0.48	85
ED	0.54	0.09	233	12	762	1.8	0.5	7	0.41	43
	(0.95)	(0.10)	(217)	(11)	(1243)	(3.0)	(0.9)	(7)	(0.25)	(33)

concentration of nitrites. An acidic milieu favours this bioconversion. When nitrites are present, haemoglobin can be oxidized to methaemoglobin, which cannot carry oxygen and causes methaemoglobinaemia. This is very dangerous for children. In a weakly acidic milieu, nitrites are converted into nitrosamines, which have been found to be carcinogenic (Klein, 1973; Musibono, 1992; WHO, 2000).

Minor fertilizer elements in the irrigation water. Calcium (Ca), magnesium (Mg), sulphur (S) and sodium (Na) are necessary macronutrients for crops. Irrigation water can also carry these elements in significant quantities. The concentrations we found in our study of these elements in irrigation water are given in Table 8.5. Concentrations of Ca and Mg were low in both seasons, although Mg levels were somewhat lower during the rainy season. The Ca and Mg values indicated that the water was soft.

Heavy metals in the irrigation water. Several heavy metals are found in wastewater contaminated by industrial pollution. Some of the more commonly found heavy metals include Pb, Cd, chromium (Cr), cobalt (Co), nickel (Ni), copper (Cu), zinc (Zn) and molybdenum (Mo). The average concentrations of Pb, Cd, Cr, Co, Ni and Cu are recorded in Table 8.6. The concentrations of Pb and Cd were generally higher in the rainy season than in the dry season. The concentrations of Pb for all samples were less than 5 mg/l,

Table 8.6 *Mean concentrations of heavy metals in the irrigation water during the rainy (dry) season, 2005*

Sampling site	Pb^{2+} (μg/l)	Cd^{2+} (μg/l)	Cr^{2+} (mg/l)	Co^{2+} (mg/l)	Ni^{2+} (mg/l)	Cu^{2+} (mg/l)
ENDA	264	122	0.33	0.09	0.22	1.8
	(239)	(132)	(0.26)	(0.07)	(0.31)	(1.6)
ENDM	347	184	0.26	0.10	0.29	1.5
	(162)	(108)	(0.24)	(0.06)	(0.27)	(1.7)
EPP	517	155	0.24	0.03	0.32	2.0
	(191)	(117)	(0.25)	(0.03)	(0.29)	(1.3)
EET	261	157	0.22	0.08	0.303	1.9
	(243)	(143)	(0.25)	(0.04)	(0.32)	(1.6)
EMA	532	256	0.26	0.09	0.35	1.7
	(265)	(168)	(0.25)	(0.12)	(0.41)	(1.7)
EPA	279	276	0.21	0.04	0.30	2.3
	(182)	(109)	(0.26)	(0.04)	(0.43)	(1.7)
EPNA	411	387	0.24	0.09	0.32	2.2
	(182)	(103)	(0.25)	(0.08)	(0.32)	(1.6)
ERO	418	317	0.23	0.04	0.28	1.6
	(200)	(163)	(0.25)	(0.05)	(0.33)	(1.5)
EJE	631	286	0.25	0.05	0.49	1.5
ED	473	209	0.33	0.07	0.31	1.5
	(217)	(118)	(0.28)	(0.07)	(0.27)	(1.5)
FAO standards	5000–10,000	10–50	0.1–1.0	0.05–5.0	NA	0.2–5.0

the maximum authorized value for long-term use of irrigation water (FAO, 2003); whereas the concentrations of Cd^{2+} were, in most cases, more than ten times higher than the allowable limit. In about half the samples, concentrations of Co were less than 0.05 mg/l, allowable for long-term use. Ni values were higher than the allowable value of 0.2 mg/l. In a couple of sampling locations, Cu values were higher than the allowable value of 2.0 mg/l. Heavy metals can be a significant risk for humans and animals (FAO, 2003); moreover, the bioaccumulation of cadmium can be toxic in humans (FAO, 2003), and high levels of Zn, Cu and Ni can kill plants.

Bacteriological quality of irrigation water and potential risk of human contamination. Contaminated wastewater used for irrigation poses a risk to vegetables, farmers and consumers (Mara and Cairncross, 1991; Niang, 1996). This risk was evaluated by sampling the irrigation water for total coliform, fecal coliform, fecal streptococci and *E. coli* (Table 8.7). In the majority of cases, the water was found to be contaminated by micro-organisms, whose concentrations were higher than acceptable standards.

Farmers working in market gardens are exposed to several risks, most notably exposure to *E. coli* and *Salmonella dysenteriae* in the water; insects such as mosquitoes, tsetse flies and bees; and metallic objects in household garbage. The majority of market gardeners do not use protection such as

Table 8.7 *Bacteriological quality of irrigation water, 2005*

Sampling site	Most probable number/100 ml			
	TC	FC	FS	EC
ENDA	>5500	13.1	+	+
EPP	400–6000	6.5	+	+
EET	550–5500	6.5	+	+
EPA	67–667	1.1	+	+
ERO	50–500	1.1	+	+
EPNA	550–5500	7.2	+	+
EMA	>4000	11.5	+	+
FAO standards		≤1000		

Each value is the average of six analyses. EC, *Escherichia coli*; FC, fecal coliform; FS, fecal streptococci; TC, total coliform.

boots and gloves while working. The main illnesses affecting market gardeners are malaria (39.6 per cent), typhoid fever (18.81 per cent), rheumatism (17.49 per cent), gastroenteritis (9.9 per cent), amoebiasis (8.9 per cent), haemorrhoids (5.9 per cent), itchy skin (4.3 per cent), general aches and pains (2.9 per cent), muscle pain (1.3 per cent), sinusitis (1.3 per cent) and schistosomiasis (0.3 per cent). A significant number of the market gardeners were unaware sickness can be transmitted by irrigation water. This lack of knowledge was found to be almost identical for men (35.4 per cent) and women (32.8 per cent).

Effect of Projet de Compostière de Quartier compost on yield

During the period of research from October 2004 to September 2005, there were four crop rotations, with the land lying fallow during May, June and July. The time lapse between transplanting and harvest took 25–30 days. Table 8.8 shows the size and mass of vegetables and length of leaves in plots with and without PCQ for the first three rotations. Each value is the average of ten specimens.

The results were mixed. The size of the vegetables and the length of their leaves were greater with the use of sludge. In contrast, values for these parameters decreased with the use of the PCQ compost. In general, irrigation with wastewater resulted in increased vegetable and leaf size than irrigation with tap water. This difference is explained by the greater amount of nutrients in the wastewater which stimulates growth. The mass of vegetables grown with PCQ compost diminished over the rotations, but there was a remarkable improvement over those grown with sludge. Findings indicate that supplementary PCQ compost should be added after the second rotation. However, it seems that sludge does not begin to produce these effects until after the third rotation (150 days).

Table 8.8 *Parameters associated with experimental garden vegetables*

Flower beds	Parameters	Rotation		
		1	2	3
PCQ–WW	SGV (cm)	70	56	36
	LGV (cm)	23	18	14
	MG (kg)	45	36	22
PCQ–FW	SGV (cm)	38	32	25
	LGV (cm)	15	16	12
	MG (kg)	24	29	11
SLU–WW	SGV (cm)	20	47	40
	LGV (cm)	11	16	19
	MG (kg)	9	13	16
SLU–FW	SGV (cm)	18	40	44
	LGV (m)	10	31	38
	MG (kg)	10	12	15

PCQ, compost; SLU, sludge; WW, wastewater; FW, faucet water; SGV, size of experimental garden vegetables; LGV, leaf length of experimental garden vegetables; MG, mass of vegetables harvested per standard plant bed ($12\,m^2$).

CONCLUSION AND RECOMMENDATIONS

The characteristics of UA in Kinshasa in general, and in the Centre Agricole de N'djili in particular, were described and analysed to identify constraints on UA and solutions to promote safe UA. Physical, chemical and bacteriological characteristics of the water used to irrigate the market gardens were also determined.

Market gardeners in Kinshasa are grouped in cooperatives distributed over 32 agricultural centres and they tend to live below the poverty threshold, although most have some education. Market gardening is an activity that ensures food security for the citizens and allows the market gardeners to generate some income. Market gardening helps to reuse wastes: both solid organic matter through compost and liquid wastes through the application of wastewater. While this nutrient recycling is positive, due to pervasive contaminants in the soils and irrigation water, UA can also have a negative impact on the environment, on farmers themselves and those that consume the products.

This study shows that market gardeners of the Centre Agricole de N'djili were exposed to microbial and chemical contamination from the irrigation water. They engaged in risky behaviour either due to a lack of understanding of the risks or because they lacked adequate means to protect themselves. Farmers, however, did not see health as a priority issue for themselves, but instead highlighted the lack of seeds, lack of fertilizer and lack of financial support as the main constraints to their livelihood.

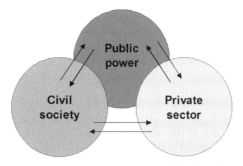

Figure 8.1 *Actor interactions in a sustainable development and good governance context*

In this agricultural centre, the research showed that water is polluted in part because of the rampant and unregulated use of chemical fertilizers and pesticides. These problems, however, are surmountable but need better strategies for urban management. With regards to the Centre Agricole de N'djili, solutions are summarized in a 'Sustainable Environmental Governance' model that outlines three complementary and closely overlapping groups – public power, civil society and the private sector (Figure 8.1).

Further research is needed to determine the source of chemical and bacteriological contamination of the market gardeners' water. Kinshasa also suffers from the lack of laboratory facilities to test the biological and chemical qualities of the water used in market gardening which should become an important objective of new investments in research.

ACKNOWLEDGEMENTS

This study was made possible by financial support from IDRC's International Graduate Research Awards in Urban Agriculture (AGROPOLIS), which allowed me to conduct this research and obtain necessary materials. I also thank Professor Noki Vesitiluta and Professor Musibono Eyul'Anki, advisor and co-advisor for the doctoral thesis, respectively. Thanks are also given to the market gardeners of the Centre Agricole de N'djili; to the Union of Market Garden Cooperatives of Kinshasa (UCOOPMAKIN) and the Coopérative Agricole du Centre de N'djili (COPACEN) management committees; to the researchers involved in this study; and to Nelly Nzuzi Panzu, the laboratory technician.

NOTES

1 Kifuani Kia Mayeko, Ph.D. (candidate) Environmental Management; Mvuzi Street, 3036, Q. Gombele, Lemba, Kinshasa, DRC; tel: +243 09 98 22 99 87; email: kifuani@ yahoo.fr
2 The remaining 11 per cent did not respond to the question.

REFERENCES

Agripromo (1985) 'Le jardin familial', *Agripromo*, vol 51

Alexeev, V. (1980) *Analyse Quantitative*, 3rd edition, Mir, Moscow, Russia

Arrignon, J. (1991) *Aménagement Piscicole des Eaux Douces*, 4th edition, Techniques et Documentation, Lavoisier, Paris, France

Autissier, V. (1994) *Jardins des Villes, Jardins des Champs: Maraîchage en Afrique de l'Ouest, du Diagnostic à l'Intervention*, Collection le point, Gret, Paris, France

Ayres, R. M. and Mara, D. D. (1997) *Analyse des Eaux Résiduaires en Vue de leur Recyclage en Agriculture*, World Health Organization, Geneva, Switzerland

Barry, M. B. (2005) 'Determination of urban livestock adoption in the 'zone dense' of Khorogo, Côte d'Ivoire: A two-bit approach', in Mougeot, L. J. A. (ed) *Agropolis: The Social, Political and Environmental Dimensions of Urban Agriculture*, Earthscan, London, UK

Berl, W. G. (1951) *Physical Methods in Chemical Analysis*, vol 2, Academic Press, New York, NY, USA

Bontoux, J. (1983) *Introduction à l'Etude des Eaux Douces*, Éditions Cebedoc, Liège, Belgium

Boole, M. (2000) 'Microcrédit et production maraîchère à Kinshasa: cas de maraîchers de N'djili/CECOMAF', Mémoire de Licence, Faculté des Sciences Agronomiques, University of Kinshasa, Kinshasa, Democratic Republic of Congo

Crabbe, M. (1980) *Le Climat de Kinshasa d'après les Observations Centrées sur la Période de 1931–1970*, édition assurée par les services de l'administration Belge et de la coopération au développement, Brussels, Belgium

Delvaque, J. (1980) *Étude pour une Planification des Cultures Maraîchères au Sénégal vol 1: Les Facteurs de Production, leur Répartition Régionale*, Centre pour le Développement de l'Horticulture du Sénégal, Dakar, Senegal

Dongmo, T., Gockowski, J., Hernandez, S., Awono, L. D. K. and Mbang à Moudon, R. (2005) 'L'agriculture périurbaine à Yaoundé: ses rapports avec la réduction de la pauvreté, le développement économique, la conservation de la biodiversité et de l'environnement', *Tropicultura*, vol 23, no 3, pp130–135

Due, J. M. (1986) 'Agricultural policy in tropical Africa: Is a turnaround possible?', *Agricultural Economics*, vol 1, no 1, pp19–34

Eckenfelder, W. W. (1982) *Gestion des Eaux Usées Urbaines et Industrielles*, Lavoisier, Paris, France

Fall, S. T. and Fall, A. S. (ed) (2001) *Cités Horticoles en Sursis? L'Agriculture Urbaine dans les Grandes Niayes au Sénégal*, International Development Research Centre, Ottawa, Canada

FAO (2003) *Irrigation avec des Eaux Usées Traitées – Manuel d'Utilisation*, FAO, Rome, Italy

Fluoriot, J., de Maximy, R. and Pain, M. (1975) *Atlas de Kinshasa*, Bureau du Président de la République, Institut Géographique du Congo (IGC); Bureau d'études et d'aménagement urbain (BEAU); Ministère des Travaux publics et de l'Aménagement du Territoire (TRAT), Kinshasa, Democratic Republic of Congo

Hach Chemical Company (1979) *Hach Handbook of Water Analysis*, Loveland, CO, USA

Hovorka, A. J. (2005) 'Gender, commercial urban agriculture and food supply in greater Gaberone, Botswana', in Mougeot, L. J. A. (ed) *Agropolis: the Social, Political and Environmental Dimensions of Urban Agriculture*, Earthscan, London, UK

Institut National de la Statistique (INS) (1991) *Zaïre – Recensement Scientifique de la Population, Juillet 1984: Totaux Définitifs*, INS, Kinshasa, Democratic Republic of Congo

Kifuani, K. M. (2004) 'Étude de la qualité des eaux des effluents d'eaux usées domestiques', MSc thesis, Faculté des Sciences, University of Kinshasa, Kinshasa, Democratic Republic of Congo

Kiyombo, M. (2005) Manuel des travaux pratiques destiné aux etudiants de l'ecole de Santé Publique University of Kinshasa, Kinshasa, Democratic Republic of Congo

Klein, L. (1973) *River Pollution, Chemical Analysis*, 6th edition, Academic Press, New York

Lele, U. (1986) 'Women and structural transformation', *Economic Development and Cultural Change*, vol 34, no 2, pp195–221

Mara, D. and Cairncross, S. (1991) *Guide pour l'Utilisation sans Risques des Eaux Résiduaires et des Excréta en Agriculture et Aquaculture*, World Health Organization; United Nations Environment Programme, Geneva, Switzerland

Ministère du Plan (2004) *Document de Stratégie de Réduction de la Pauvreté (DSRP)*, Ministère du plan, Kinshasa, Democratic Republic of Congo

Mokili, J. (1998) *Politiques Agricoles et Promotion Rurale au Congo–Zaïre*, Éditions l'Harmattan, Paris, France

Mougeot, J. A. (ed) (2005) *Agropolis: the Social, Political and Environmental Dimensions of Urban Agriculture*, Earthscan, London, UK

Mukana, W. M. and Kifuani, K. M. (2000) 'Préparation des charbons actifs à partir des sciures de bagasse de canne à sucre, de bois Ntola et Lifaki imprégnées dans des solutions de soude caustique', *Revue Congolaise des Sciences Nucléaires*, vol 16, no 1, pp84–92

Musibono, E. (1992) *Qualité de l'Eau et Aquaculture*, Édition MTD Engineering, Kinshasa, Democratic Republic of Congo

Niang, S. (1996) 'Gestion des déchets urbains – l'utilisation des eaux usées brutes dans l'agriculture urbaine au Sénégal: Bilan et perspectives', available from www.idrc.ca/fr/ev-33719-201-1-DO_TOPIC.html (accessed on 17 October 2007)

Nisbet, M. and Verneaux, J. (1970) 'Composants chimiques des eaux courantes: discussion et propositions des classes en tant que base d'interprétation des analyses chimiques', *Annales de Limnologie*, vol 6, no 2, pp161–190

Rodier, J. (1984) *L'Analyse de l'Eau*, 7th edition, Dunod, Paris, France

Saint-Moulin, L. de (1976) *Atlas des Collectivités du Zaïre*, PUZ, Kinshasa, Democratic Republic of Congo

Symposium des Organizations et Dynamiques Maraîchères (2004) 'Rapport du symposium sur les dynamiques et organizations maraîchères de Kinshasa', 27–29 Avril 2004, UCOOPMAKIN, Kinshasa, Democratic Republic of Congo

Tallaki, K. (2005) 'The pest-control system in the market gardens of Lomé, Togo', in Mougeot, L. J. A. (ed) *Agropolis: The Social, Political and Environmental Dimensions of Urban Agriculture*, Earthscan, London, UK

Tanawa, E. and Djeuda Tchapanga, H. B. (1998) 'Gestion de l'eau et protection de la ressource', rapport final, École Nationale Supérieure Polytechnique Yaoundé, Cameroon

UNICEF (2002) *La Situation des Enfants dans le Monde*, Unicef, Geneva, Switzerland

United Nations Development Program (UNPD) (2001) 'Séminaire international pour l'élimination de la pauvreté: Gouvernance, nouvelles technologies et lutte contre la pauvreté', 31 Octobre 2001, University of Kinshasa, Kinshasa, Democratic Republic of Congo

Van Caillie, X. (1983) 'Hydrologie et érosion dans la région de Kinshasa', Ph.D. thesis, Université de Louvain, Ohain, Belgium

WHO (2000) *Directives de Qualité pour l'Eau de Boisson de l'OMS*, 2nd edition, vol 2, WHO, Geneva, Switzerland

9

The Health Impacts of Farming on Producers in Rosario, Argentina

Patricia Silvia Propersi[1]

INTRODUCTION

The health of a population is influenced by complex socio-ecological processes characterized by: feedback loops across space and time; social organization and methods of production, distribution and consumption; and the successive changes that occur as social forces reach an equilibrium. Birley and Lock (1999) reported that development projects can sometimes indirectly have both positive and negative effects on the physical and social environment as well as on human health. Projects that do not take into consideration the potential negative impacts on health can transfer hidden costs to the health sector, placing a burden on health services.

This paper reports the findings of a study that was carried out to determine the effects of farming conditions on the health of the people who live and work on the *quintas* (peri-urban vegetable farms) in Rosario. The city of Rosario has a green peri-urban belt where agriculture is common. The 2000–2001 horticultural census (FCA–UNR, 2001) recorded 194 agricultural farm units covering 3663 ha in the peri-urban area around Rosario. These were primarily family farms which employ peri-urban agriculture (UPA) to produce vegetables for their own consumption and for market. Vegetable production is done by family members who, in some instances, engage a number of *medieros* and day labourers to assist in various activities (Propersi, 1999). Much work on urban agriculture (UA) has been conducted in Rosario (see Van Veenhuizen, 2006, for an overview); however, little has been done on peri-urban farms. A review of literature found little or no health information specific to urban farmers.

Family-driven farming, mostly horticultural, represents a large and historic sector in the region. This has influenced settlement patterns (Table 9.1) whereby a significant number of people, including farm workers and their families, live on the quintas.

Table 9.1 *Distribution of population in different settlement types*

	Quinta	Town	City	SD
Producers	43	18	24	1
Producer's family	23	9	9	1
Medieros	38	14	5	2
Mediero's family	49	16	5	2
Labourers	36	31	6	12
Total	189	88	49	18
	55%	26%	14%	5%

The working conditions and environment associated with these horti-cultural production areas that either favourably or unfavourably influence the population's health were of interest in this study. In essence, the labour process, consumption patterns, lifestyle and home life all contribute to a persons 'epidemiological history' (Breilh, 1997, p100). Perceptions regarding health were also explored in this study. No epidemiological work has been done on farmers in Rosario in the past.

Access to health services was through City Hall primary care centres which were geographically closest to the quintas. There are three health centres near the horticulture zone: El Gaucho, Tío Rolo and San Vicente de Paul. These provide primary care, especially for women and children. Fewer men go to the health care centres since it is more difficult for them to get leave from the quintas where they work.

OBJECTIVES AND HYPOTHESES

General objectives

The general objective of this study was to document, characterize and understand the health risks and protective measures used in vegetable production within the horticultural units, and to assess their impact on the health of the population closely related to the production.

Specific objectives

The specific objectives were to:

1 understand and describe the socio-technical and organizational factors associated with vegetable production processes in the Rosario green belt
2 describe the characteristics of the population involved in vegetable growing and the specific links this population has with the production units
3 analyse the environmental conditions and production practices that may have negative and positive consequences for the health of the population involved
4 understand the perceptions this population had about health risks.

Hypotheses

The following three hypotheses were defined for this research:

1   A technological model increases the productivity of natural resources, which influences the organization of production activities.
2   Living in the vegetable-production unit hinders workers' access to goods and social services.
3   There is a strong correlation between growing vegetables for market and the use of toxic inputs that negatively affect the health of the population.

METHODS

Field survey

The study area was limited to the green horticultural belt of the Rosario peri-urban area, which includes Constitución, Rosario and San Lorenzo in Santa Fe province, Argentina. For this study, a quantitative methodology was used to delimit the objectives of the study and to define the variables. Statistical tests were used. Also, cross-references with additional analysis and the subsequent application of a qualitative methodology were used to decipher the findings. Interviews of the key actors within both the production units and the institutional web of health services provided qualitative information.

The database used in this study was generated from the May 2000 to June 2001 census of the area, which identified 194 farms which were focused exclusively on horticultural production (FAC–UNR, 2001). They were classified into three modal types: high risk, medium risk and low risk (Table 9.2).

Table 9.2 *Modal types for the production units*

Modal type	Number of production units
High risk	97
Medium risk	36
Low risk	61

From the information obtained by the census, modal types were selected to consider the different conditions and working environment of people living and working there. These were in turn used to develop 'risk rankings' based on the potential exposure of farmers to different variables such as pesticides, heavy machinery and fatigue. The five main elements considered were:

1   Size of the production unit.

2   Production diversity, as in the number of different species produced on the farm. The diversity and type of production suggested the quantity and quality of labour required on the farm.

3   Intensiveness of the production unit, as determined by the number of hectares farmed for each crop, the way in which crops were produced – that is, in technological and capital assets used – and the types of crops grown under any type of cover (greenhouse, tunnels, etc.). This helped establish risk ranks according to the generic or specific chemical products used to control diseases. The type of chemical inputs depended on the type of crops as well as the technology used for their production. Intensiveness also provided an indicator of time spent in a day's work.

4   Characteristics of the labour organization, incorporating details such as the number of people working and their type of jobs (owner–producer, familiar, mediero, salaried worker, day labourer). Also, labour contracts provided information on the employment structure with regard to work organization and responsibilities.

5   Residence, age and gender of the different people on the farm – family or labourer – posed the probable-risk magnitude from the environment where they live in relation to their age and their gender.

Values were assigned for each of the above indicators and a risk ranking was formed, as illustrated in Table 9.2.

The size of each production unit (in hectares) was the reference variable in a stratified sample of the 194 identified producers. Size was chosen because its absolute variability was minor compared with that of other variables. A sample size of 73 producers was determined to be representative, with a 95 per cent level of confidence.

Data for the field survey were collected from the owner–producer of each quinta. In many cases, the owner–producers did not allow other people working in the unit to participate in the survey. As a result, the only questions that could be asked were those that the owner of the production unit would answer. Although this was a constraint, the results nevertheless are still useful. Further work can build on this initial survey.

A structured questionnaire was used to obtain information from the horticultural producers. It included questions about production conditions, working conditions, the physical environment, access to basic services and personnel's health: of the owner–producer, the owner's family and any other person working or living in the horticultural production unit.

The questionnaire had the following three characteristics:

1   It exposed each interviewee to 'the same stimulus' (questions), which enabled all the data to be grouped and compared.

2   The questions had a predetermined sequence.

3   The breadth of the study was determined beforehand and lists of potential responses were prepared in advance for categorization. Some semi-structured questions were included to elicit a wide range of responses.

Statistical data

Inferential and descriptive statistical techniques and computer programs were used. The software used was the Statistical Analysis System, an integrated system of products provided by the SAS Institute for Windows Version XP, and Statistics Toolbox 6.0 (MathWorks, Inc.), a dataset used for organization and analysis of heterogeneous data and metadata.

Qualitative data

Information acquired from owners was compared with data from workers who were interviewed both inside and outside the quintas. The latter interviews enhanced the scope of the study and were used to assess risk and favourable processes as well as to determine the extent of female and child labour, which was frequently hidden in production units. In many cases, workers denied this was going on; therefore, photographs were used to prove its existence as a reality in the workplace. Health professionals were interviewed in the care centres used by farmers to cross-reference information gathered on the farms.

Production conditions on the farms as well as the impact of production on the health of the population were determined from the surveys and interviews. Attempts were made to ascertain how the actors perceived and built their social reality based on their social class.

RESEARCH FINDINGS

Structure of the local horticultural activity

The peri-urban green belt around Rosario contains 194 horticultural farms of a variety of sizes (Figure 9.1). In general, the average enterprise is substantially smaller than other agricultural and animal husbandry farms (Cloquell et al, 2007).

More than half (53 per cent) of the farms are 10 ha or less. The vegetables produced in this peri-urban area are diverse and, therefore, a variety of techniques and resources are applied for different varieties. Moreover, the diversity makes it possible to keep land in production all year round. Different crops are planted depending on seasonal cycles and crop growth time, varying from a few weeks for lettuce and radishes to years for asparagus and artichokes. Other production factors are also influential. For example, the production of tomatoes and peppers are capital intensive and require high-quality labour, in contrast to the production of carrots and marrows.

In intensive production such as horticulture, the relationship between labour and capital is connected by the type of technology used during the production process. Labour inputs can be characterized differently in horticultural production depending on resources, size and composition of the family unit, production strategy, marketing and so on. Four different

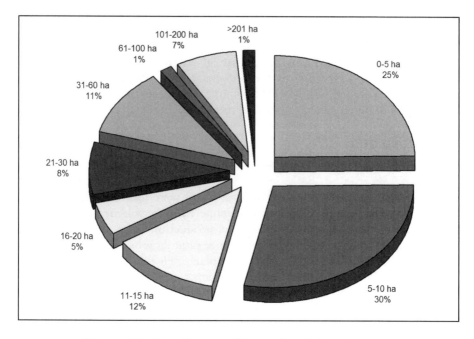

Figure 9.1 *Distribution of horticultural farms by area*

categories of actors involved in performing farming activities were identified:

1   Owner–producers work in production and marketing and includes labour
    performed by family members either on a part- or full-time basis.
2   Medieros are responsible for the physical work needed to produce crops.
    They may also call on family labour, to avoid the costs of hiring salaried
    workers during the harvest. Photographs and information given by reliable
    informants show that work done by women and young children makes an

Table 9.3 *Composition of groups of workers on the quintas*

Horticultural farms by area (hectares)	Total	%
0–5 ha	18	24.66
5–10 ha	21	28.77
11–15 ha	9	12.33
16–20 ha	4	5.48
21–30 ha	6	8.22
31–60 ha	8	10.96
61–100 ha	1	1.37
101–200 ha	5	6.85
More than 201 ha	1	1.37
Total	73	100.00

important contribution. This group comprises migrant workers from inside the country and from neighbouring countries, especially Bolivia.

3   Permanent, salaried workers usually work under oral agreement and are not bound by (formal) contracts.

4   *Tanteros*, also known as *jornaleros*, are temporary salaried workers who combine tasks on the farm with other jobs such as bricklaying or transport in Rosario.

The number of workers in each quinta varied from 1 to 12, with an average of four. Also, a significant number of people lived on the quintas.

Women and children in the workforce

In general, when enquiries were made about labour, respondents only made reference to work done by men and gave no account of the work women and children perform. However, interviews exposed some of the more specific gender-related information:

> *My wife? I would say she works more than me, because she comes back to the house at 11:00 h to start cooking, cleaning, doing the laundry and also to take care of the children. As soon as she finishes eating, she does the washing up. I stop working at 12 and come back to work in the quinta at 14:00 h, in that time she is cleaning and cooking. After doing the housework, she is back in the quinta to work. My wife knows she has things to do in the house and she takes no break. (Mediero, Santos)*

> *The children of the medieros also work in the units because some tasks are more suitable for a child than for an adult. Not all of them work, of course, and its not that they are going to grab an escardillo (a heavy tool). I mean, children take care of the hoeing, cleaning and the like. That kind of work is better done by children because of their size: they don't need to bend to do it! Maybe with the years they start feeling the consequences. (Horticultural worker, Fernando)*

The reported instances of work performed by children and women were confirmed during interviews with doctors in the health care unit, as well as by checking medical records that indicate recurring minor injuries in children. When doctors repeatedly ask about the cause of the injuries, children's work is finally mentioned. However, a paediatrician at the Tio Rolo Health Care Unit reported that unless medical staff press for details, parents do not admit their children are working.

The labour environment

The peri-urban area where farms are located is defined by the municipality of Rosario as a protected green belt that is closed to urbanization

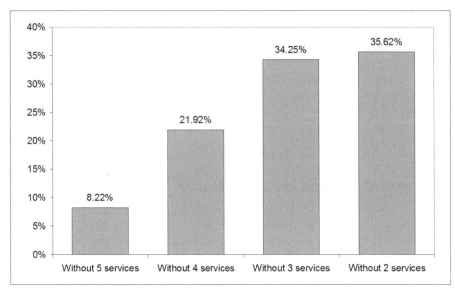

Figure 9.2 *Access to services where workers reside*

('non-urbanizable' is the term used by the city). Infrastructure of basic services such as electricity, water supply, garbage collection, asphalt roads, mass transport, gas, sewers and telephones, among other things, either do not exist or are limited, all of which leads to difficulties in accessing public services (Figure 9.2). This study focused on the green belt, where urban development is not permitted.

Some peri-urban areas where horticultural production units were found were more than 3 km from workers' homes, being well outside the city limits. Such areas represented a geographical barrier for the farm population, who often had to return home in the dark (Díaz et al, 2000, p13). Moreover, only 23 quintas (31.5 per cent) offered a bus service.

Studies by the Food Institute of Rosario (part of the municipal health secretariat at Rosario City Hall) that have assessed the quality of drinking water have found that 59 per cent of the samples had fecal coliform counts which exceeded acceptable norms for drinking water. In addition, *E. coli* were present in 31 per cent of the samples (Llanes et al, 2006). However, the survey showed that less than half the population is concerned about the quality of the drinking water. In contrast, water for irrigation is more closely monitored. This indicates that despite the presence of services on farms, there are health risks associated with water provision on them.

Duration of the working day

The care of crops and field work is largely done by *medieros*, with commercial and administrative work being done by the owners of the quinta.

Table 9.4 *Hours worked on average during the different seasons*

Season	Working hours (mean)
Summer	10.55
Autumn	8.4
Winter	7.61
Spring	9.5

Working conditions are not regulated and there is almost no presence or monitoring by the government. Instead, the duration of the working time and hours are determined by the needs of the different crops (Table 9.4). Answers recorded in the surveys regarding hours worked during the different seasons show that significantly more are required in spring and summer compared with other times of production.

The preparation of the soil is one of the most repetitive tasks of the productive process. Fifty-nine percent of the work done by quinta owners involves the use of a tractor and machinery, while other tasks require manual control of weeds. In-depth interviews and photographic records showed instances of soil-removal activities performed by contracted workers using horse-drawn machinery. In larger quintas, 70 per cent of the manual tasks are performed by contracted workers, either as employees or as *medieros*. Not surprisingly, commonly reported problems associated with farming amongst men in the health-care units included chronic problems associated with back, leg, neck and arm pain.

Processes involving handling of chemical products

Intensive horticulture often requires agri-inputs such as fungicides, insecticides and manure. Use of these products can cause a variety of health and environmental risks if applied inappropriately. Often, mishandling is associated with a lack of understanding of instructions such as application interval charts for chemical products.

Spraying using a backpack comprised one common application method of pesticides that was reported by 42 per cent of the respondents. Studies on the use of backpacks for manual spraying of pesticides (Table 9.5) showed that parts of the operator's body are exposed to a great deal of residue (Bulacio

Table 9.5 *Method used to apply pesticides*

Backpack sprayer (%)	Fumigation machine (%)	Both (%)	Neither (%)
42.5	27.3	28.8	1.4

et al, 2001), with higher levels of chemical concentration showing on the feet, legs and thighs.

Interviews with workers graphically indicated how children become contaminated with chemical residues:

> *Picture a little boy that is in his house while his father has gone to work in the quinta for eight hours. When the father comes back home, what's the first thing his son does? The child runs to hug his father. And to which part of the father's body the boy gets hold? The legs, obviously, because they are at the little boy's height. The thing is, the legs are one of the parts of the body that have more contact with the chemical products, because it is the legs that have more contact with plants than other parts of the body like the shoulders or arms. It is the trousers that get dirtiest and most contaminated with chemicals. (Agricultural engineer, Luis Carrancio, INTA Oliveros)*

Table 9.6 indicates that, when chemical products are used, containers are sometimes reused without being carefully cleaned. Similar to the results associated with the general handling of chemical products for agricultural use, the workers' different levels of education do not indicate differences in their use of these containers.

Table 9.6 *Treatment of containers of chemical products related to owners' educational level (percentage)*

	Educational level						
	None	Primary	P. inc.	High school	Higher education	University	Total
Clean	15.38	38.46	7.69	7.69	7.69	23.08	100
Do not clean	3.33	58.33	0.00	31.67	1.67	5.00	100

P. inc., primary incomplete.

According to the records maintained by health-care institutions reviewed during this research, entries related to health problems associated with the use of chemical products existed but were not common. According to doctors at the El Gaucho health centre, this is because they only manifested as serious health problems later in life, when cases were handled at the hospitals. While health problems related to agro-toxicity were not found to be acute, low concentrations of these toxins were not easily detected. Therefore, their impact is 'invisible' in the affected population and were not recorded at health centres:

> *The symptoms are confused with tiredness and symptoms of being exhausted. After carrying a 20-kg backpack for eight hours, it's*

normal to have a headache, even with no chemical products involved. Or it is normal feeling a pain in the back. Same as not feeling well. Because of the poor living conditions and bad diet, how can one say which symptoms are related to chemicals and which to being tired or not eating well enough? (Agricultural engineer, Luis Carrancio, INTA Oliveros)

Health risks linked to horticultural production

Respondents' perceptions regarding problems associated with growing vegetables varied according to the positions they held on the farm. For instance, owners and workers perceived problems differently. Respondents attached different meanings to health problems they were afflicted with and set forth different strategies for solving them. It was evident that in solving health problems priority was given to issues that interfered with work and everyday life.

While serious problems are dealt with in the hospital, health staff pointed out that the workers in the quintas did not seek medical attention very often, partly because of their concept of health and illness. Workers considered getting treatment for an illness only when they experienced disability or limited movement that interfered with their work. The reluctance to seek help made it difficult for health care workers to make any progress against preventable diseases. In fact, the lack of information on such problems prevents the exposure of factors related to the overexploitation of workers, the overuse of chemicals in order to produce 'perfect' produce and unnecessary environmental deterioration of soil and water. This situation is exacerbated in the absence of governmental regulations.

When workers are asked about their existing health problems or whether they had experienced any the previous year, the answer tended to be 'nothing, luckily enough' or 'none, thank God'. However, they then continued to describe recent hospitalization for themselves or somebody else and avoided discussing major health challenges such as hypertension, which requires daily medication. Also, during interviews, respondents willingly presented their neighbours' surgical history, injuries and other illnesses. A review of clinical data corroborated these findings.

CONCLUSION AND RECOMMENDATIONS

This research was conducted with the support of the Rosario City Hall and the National University of Rosario. The arrangement between a public educational institution and the city authority in charge of formulating and executing policies is a novel approach that provided benefits to all involved. The research was designed as an introduction to some of the factors that influence health risks and exposure on farms in peri-urban Rosario. Through interviews with medical professionals, farmers and labourers, several variables, such as physical ailments, pesticide use and the relationship between exposure and labour type, have been introduced. Hopefully this will set the stage for further work on the link between peri-urban farming enterprises and the health risks implicit in such work.

Labourers on horticultural production farms in the green belt of Rosario are rarely noticed in society. As such, health risks and dangers associated with working on peri-urban farms escape scrutiny. An example cited in this work is the inappropriate use of pesticides, suggesting that awareness-raising campaigns are needed on the topic of their safe application. Civil society organizations in Rosario are well placed to be active in these campaigns; however, appropriate regulations by local authorities are necessary as well.

To affect some changes, communication strategies should be designed in partnership with educational and technology institutes at the university level, as well as any primary care centres which are highly integrated with the community. It would be helpful to review extension programmes in horticulture and to define a new strategy to increase both the quantity and quality of produce so as to sustain or develop economic growth. This is particularly relevant considering the direct effects food production (and food itself) has on quality of life. For example, training in organic production methods could be done in association with increasing knowledge on the appropriate use of agricultural inputs. The labour of women and children is also often hidden in UPA and this can lead to increased exposure to on-farm risks.

Working conditions on peri-urban farms do not comprise the worst employment environment in the area. However, actually procuring more accurate, detailed information on agricultural workers' working and living conditions means that more appropriate interventions urgently need to be designed by the authorities and organizations involved in UPA around Rosario. The relationship between the organization of agricultural production and human health represents crucial knowledge which should be incorporated into the design and implementation of appropriate agricultural extension programmes.

ACKNOWLEDGEMENTS

This paper is based on the results obtained from Ph.D. research funded by the AGROPOLIS award programme of Canada's International Development Research Centre (IDRC). In particular, I want to gratefully acknowledge the support of IDRC, especially Wendy Storey, who was always available to answer my questions. I would like to also thank my thesis supervisor, Silvia Cloquell, and Roxana Albanesi, who helped me focus my work. The personnel of the primary care centres of the Health Secretary of Rosario provided valuable support. Last, but not least, special thanks go to Jaime Breilh, Arturo Campaña, and the entire team of the Health Research and Advisory Centre in Quito, Ecuador, for their help in broadening my experience and knowledge of epidemiology.

NOTE

1   Patricia Silvia Propersi, Ph.D. candidate; Facultad de Ciencias Agrarias de la Universidad Nacional de Rosario; tel: + 54 341 4329061; email: pproper@unr.edu.ar

REFERENCES

Birley, M. H. and Lock, K. (1999) *The Health Impacts of Peri-Urban Natural Resource Development*, International Centre for Health Impact Assessment, Liverpool School of Tropical Medicine, Liverpool, UK
Breilh, J. (1997) 'Nuevos conceptos y técnicas de investigación – Guía pedagógica para un taller de metodología: Epidemiología del trabajo', *Serie Epidemiología Crítica*, no 3, Centro de Estudios y Asesoría en Salud, Quito, Ecuador
Breilh, J. (2003) *Epidemiología Críticae*, Lugar Editorial, Quito, Ecuador
Cloquell, S. (coordinadora), Albanesi, R., Propersi, P., Preda, G. and De Nicola, M. (2007) *Familias Rurales: El Fin de Una Historia en el Inicio de Una Nueva Agricultura*, Homo Sapiens Editora, Rosario, Argentina
Díaz, A., Huerta, A., Rodríguez, A. and Telesco, C. (2000) 'La dimensión sociocultural y su relación con los patrones de utilización: Acerca de un centro de salud de atención primaria desde la perspectiva de la población', *Investigación en Salud*, vol 1–2
Facultad de Ciencias Agrarias de la Universidad Nacional de Rosario (FCA–UNR) (2001) *Base de Datos del Censo Hortícola Realizado por Integrantes del Proyecto Hortícola de Rosario (campaña 2000/2001)*, FCA–UNR, Zavalla Town, Santa Fe, Argentina
Llanes, R., Maniscalco, M., Rodil, C., Casella, G. and Viñas, C. (2006) 'Análisis de establecimientos frutihortícolas', Revista: Énfasis Alimentación Latinoamérica. AñoXII-Febrero/Marzo, p42
Panelo, M. S., Bulacio, L. G., Giuliani, S. L. and Giolito, I. (2005) 'Riesgo de contaminación personal en la aplicación de productos fitosanitarios en cultivos hortícolas de diferente prote', in Bogliani, M. and Hilbert, J. (eds) *Aplicar eficientemente los agroquímicos*, Castelar, Ediciones INTA, pp246–250
Propersi, P. (1999) 'Un espacio de silencios: el mediero hortícola', in Albanesi, R., Cassinera, A., Propersi, P., Qüesta, T. and Rosenstein, S. (eds) *La Horticultura Rosarina: Comercialización, Organización Laboral y Adopción Tecnológica*, UNR Editora, Rosario, Santa Fe, Argentina
Van Veenhuizen, R. (ed) (2006) *Cities Farming for the Future: Urban Agriculture for Green and Productive Cities*, RUAF Foundation and the International Institute for Rural Reconstruction, Manila, Philippines

10

Using Participatory Education and Action Research for Health Risk Reduction Amongst Farmers in Dakar, Senegal

Nita Chaudhuri[1]

INTRODUCTION

Senegal, a country of 11.4 million people which is among the poorest in the world, is experiencing a rapid urbanization process, partly due to severe drought in the hinterland. The capital city of Dakar has a population of 2.3 million inhabitants with an urban growth of 4 per cent per annum (Editions, 2000; UNDP, 2005). Among Dakar residents, lack of formal employment, deterioration of existing urban infrastructure and lack of access to land, finance and adequate shelter (UNCHS, 2001) have led people to urban agriculture (UA) for both financial and food security reasons. Yet exploitation of the Niayes (Figure 10.1), part of a productive green belt of land that stretches through Dakar up the coast, has generated significant environmental degradation, including soil infertility, deforestation, scarcity and salinity of water, loss of biodiversity and waste accumulation. Farmers have responded to problems of soil infertility with applications of wastewater, mineral and organic fertilizers and also use pesticides to maximize production (Niang and Gaye, 2004). The unique ecosystem of the Niayes is composed of sandy dunes and depressions that are regularly flooded with groundwater from the very high water table. The quantity and quality of groundwater is vulnerable in this region because sandy mineral soils do not filter potential contaminants such as pesticides and nitrates. In addition, the excessive pumping of groundwater has caused saline water intrusion to occur, which has degraded both groundwater and soil quality (Figure 10.2) (Gueye, 2005). Intense human activity due to increasing population pressure has also severely affected the

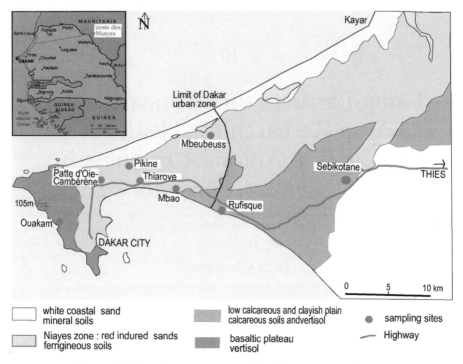

Figure 10.1 *The Niayes area of Dakar: Main soil types*

surrounding ecosystem. Options for minimizing the health impacts of the various practices that damage the ecosystem are available, but not adequately implemented. These practices include such interventions as wastewater treatment, various irrigation techniques or integrated pest management (IPM).

The doctoral research reported here complements research on wastewater treatment and ecosystem health effects undertaken by an interdisciplinary team of Senegalese researchers from the University of Dakar. It aims to demonstrate the applicability and feasibility of a process of participatory education and action research (PEAR) in response to the challenge of environmental degradation experienced by a group of farmers in Pikine, Dakar.

The chapter begins with a brief description of the theoretical underpinnings of eco-health as well as PEAR frameworks. The introduction of key objectives is followed by a brief overview of the methods used. The section focusing on results highlights three areas: farmers' perceptions and practices in relation to environmental problems and health risks; the PEAR process; and preliminary impacts of this process. In the Conclusions section, we allude to the contributions of this research and suggest future research directions.

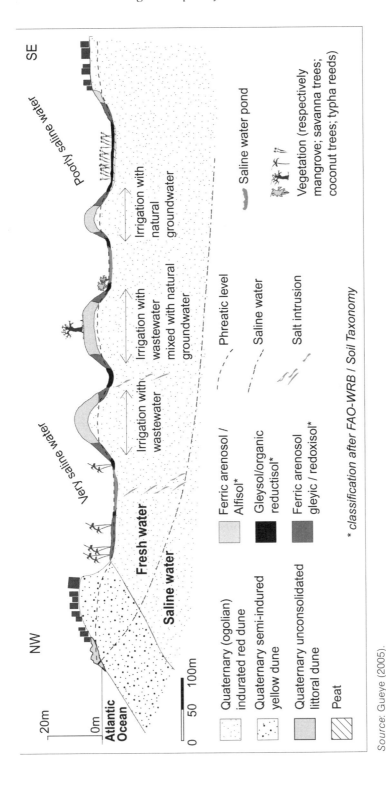

Figure 10.2 *Cross-section through the Niayes Zone in Pikine*

Source: Gueye (2005).

OBJECTIVES AND HYPOTHESIS

Agro-ecosystems such as those found in UA are complex systems in which people have deliberately selected crops and livestock to replace natural flora and fauna (Mooney et al, 1994; Alkora et al, 2004). Human health is integrally connected to such systems. The application of an eco-health framework requires the use of multiple research and action strategies applied to three methodological pillars: trans-disciplinarily, equity and participation (Lebel, 2003).

The choice to use a participatory approach is rooted in the notion that those who are affected by issues should be part of the process of defining not only the problems, but also their solutions. Participation is not only 'fair', but also provides a more thorough understanding of a situation. Positivist 'scientism', on the other hand, has a history of less participation by the subjects of research and makes meaningful participation in discourse and informed decision-making in everyday life difficult (Foucault, 1980, cited by Sohng, 1995). In an ideal world, encouraging investigation, education and action in participatory research empowers people to gain access to knowledge by co-creating and using it. Critical examination of participatory processes explores the role of popular knowledge and conventional science, issues of power, applicability of different methods and rigour in the participatory research process.

Hence, the overarching hypothesis was that 'a PEAR approach can be used to confront ecosystem health challenges with success varying with observable barriers and facilitators'. With this working hypothesis, the research sought to meet at least three research objectives:

1   Identify and document existing and perceived risks related to health and the environment in UA.
2   Identify critical points of intervention on environmental health related to UA within an eco-health framework.
3   Work with appropriate target groups, communicators, messages and techniques to bring about change through a PEAR process.

METHODS

PEAR requires the use of interactive and flexible methods, with the researcher facilitating, engaging in dialogue and stimulating action. Quantitative and qualitative data collection and various education techniques were used to: facilitate an iterative process of listening carefully to people; ask questions using a systematic process in order to ascertain how and why these problems fit into the context; synthesize key moments of understanding problems which were presented; and discuss strategies with the group in order to address problems. Among the key methods used were: meetings, focus groups, mapping, observation, questionnaires and workshops.

At the core of the process was a partnership between myself, as the lead PEAR researcher–educator, and three farmers who belonged to the executive board of the GIE Provania, a small farmers' association in Pikine founded in 1993. This association is a politically motivated body which was originally founded by farmers to address issues of land-tenure security. Its core mission has been to develop the cultivation and marketing of local agricultural products by building commercial networks. The association also works to create partnerships with aid agencies and other organizations and takes part in lobbying activities to promote the institutionalization of UA (IDRC, 2006). The association has 115 male and female members and is led by a president, secretary general, treasurer and financial officer. There is also a general assembly as well as regular meetings, which give it a legal structure that enables it to apply for funding and carry out projects.

Several institutions have been active in the zone such as the University of Dakar, ENDA Rup (an environmental and development NGO), Agence Nationale de Conseils Agricoles Rurale (ANCAR) and the Institut Africain de Gestion Urbaine (IAGU). They have been particularly involved in wastewater treatment, effective irrigation practices, composting and in the overall improvement of farming practices. Partnership with these institutions was essential for the PEAR process, as these organizations provided initial training and access as well as interpretation of the research context.

Other factors affected the research. For example, because the researcher was unable to speak the local language, Wolof, it was impossible to converse with the most marginalized, illiterate farmers. Moreover, the researcher's only interaction was with men. Despite women's crucial roles in marketing produce and notwithstanding their direct exposure to pesticides and wastewater, the researcher was unable to work with women. It may have been possible to work with these two core groups had she been able to work in small groups with the help of a translator.

There were four distinct phases of the PEAR process which consisted of onsite and offsite planning and development. Six field trips were made between 2004 and 2006, where a variety of interventions were developed including focus groups, questionnaires, workshops, mapping exercises, scenario drawings and the identification of risk practices. Ongoing contact via e-mail and telephone enabled continued planning and development over the entire process. The four phases of PEAR were:

1   During 2004–2005, information was gathered and disseminated by way of meetings, focus groups, in-depth interviews, mapping exercises, seasonal calendar exercises and questionnaires on UA, wastewater and pesticides. The researcher worked in collaboration with experts to design and develop materials for workshops which were given on such topics as UA, pesticides, key environment and health issues and wastewater. I then integrated this work and redesigned workshops with farmers which subsequently were conducted in the fields.

2   In 2005, questionnaire and focus group data were analysed in several ways: by door-to-door visits made by farmers, where they asked questions of

other farmers; at regular monthly meetings held by farmer's associations to discuss health and environment issues; via IPM and other expert-led workshops; and by videos made of the field site.

3   In 2006, the researcher returned to the field to develop work plans, implement evaluation questionnaires, and develop risk-practice messages and agricultural risk-practice scenario drawings. In addition, farmer leaders conducted workshops to evaluate the farmers' understanding. Workshops included presentations of new farmer-conceptualized maps, results of questionnaires and body mapping (showing health impacts). In addition, 'factors, barriers and solutions' exercises were conducted to develop action plans to protect against environment and health risks, as well as to carry out structured observation of the site.

4   After evaluating the results, feedback was presented to the farmers. A total 154 farmers were involved in the process.

Meetings

Regular meetings with GIE Provania and other farmers provided a forum for listening, dialogue and planning for education and action. They also provided an opportunity for the research team to socialize with farmers. Meetings took place in the field of the president of GIE Provania, in a straw shelter close to a set of wastewater treatment tanks.

Focus groups

Focus groups were conducted throughout the research process to identify key issues and monitor and evaluate the PEAR process (Figure 10.3). Farmers were asked a set of questions, initially derived from the literature and based on observation, which explored key constraints linked to UA and environmental health impacts. Notes were taken during the focus groups and coded for themes. These were then presented to a core group of farmers so as to validate the interpretation of responses and provide a basis for the design of interventions. Senegalese extension workers, students and researchers frequently helped with translation and interpretation as sessions were held in French and most farmers spoke Wolof.

Mapping

Mapping is a commonly used tool in water and sanitation PEAR, as it enables participants to visualize their surroundings, identifying key points of interest and linkages between environment and health. In January 2005, mapping was used to share knowledge regarding the study site. With the help of a research colleague (M. Guisse, 2005, in-depth interview) from the Institut Fondamental de l'Afrique Noire (IFAN) at the University of Dakar, who provided translation into Wolof (the national language), ten farmers were asked to use flip charts to draw key features of the zone. The mapping exercise was

Figure 10.3 *Focus group meetings with farmers in Pikine, 2005*

repeated in March 2006 during the evaluation phase of the research with the same ten farmers (Figure 10.5).

Observation

Observations of interactions and site conditions were recorded on an ongoing basis in field notes, supplemented periodically by photographs and video recordings. Specific observation times were also set aside for transect walks through the farming site. Field notes were transcribed and coded for themes that emerged during the research process. N-vivo software was used to manage the rich qualitative data from both the field notes and the focus groups.

Questionnaires

During the research process, four separate questionnaires regarding different and at times overlapping samples were administered by farmers, the researcher and university students to a total of 154 farmers. These individuals were selected by random sampling and were between the ages of 25 and 50, with 150 out of 154 being men. The preponderance of men in the sample is due to the fact that most women involved in UA are active in the markets; therefore, they are not present in the fields. The goal was to gather information on perceptions of health and environment risks as well as practices across a broad set of farmers, not just those who participated in the focus groups.

These questionnaires were developed in collaboration with farmers, the researcher's supervisors and other researchers at the University of Dakar and the University of Lausanne. Reliability of the questionnaires was assessed by having the same ones implemented twice, first by a farmer and subsequently by a university student. The way in which questions were asked in Wolof and the way responses were given in these questionnaires were compared and few important differences were found. Descriptive summaries of questionnaire data were provided to the farmers during feedback sessions and workshops.

Workshops

Interactive workshops on wastewater and pesticide health and environment effects, and methods to reduce their negative impacts, were an important method for training and issue identification. Experts from the University of Dakar, ANCAR and others facilitated initial training workshops, which were then replicated throughout the zone by trained farmers. Figure 10.4 illustrates

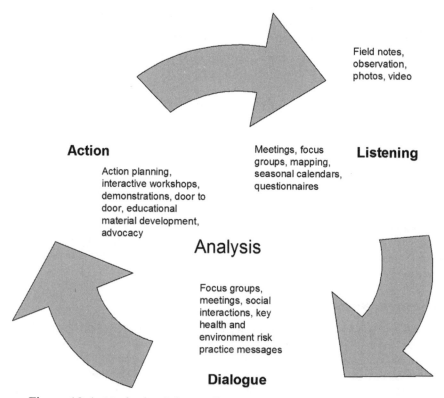

Field notes, observation, photos, video

Action

Action planning, interactive workshops, demonstrations, door to door, educational material development, advocacy

Meetings, focus groups, mapping, seasonal calendars, questionnaires

Listening

Analysis

Focus groups, meetings, social interactions, key health and environment risk practice messages

Dialogue

Figure 10.4 *Methods of data collection, participatory education and action research*

the iterative use of methods and the role they played in listening, dialogue and action, as well as in overall analysis and problem solving.

Critics of PEAR suggest that bias arises when researchers come in close contact with and empathize with individuals involved in research, thus losing their supposed objectivity. An alternative view is that data are co-constructed by the researcher and participants while situations change when we are interacting with one other and our environments. In other words, reflection is at the heart of socialized knowledge. Human values enter into the research process during problem selection, instrument design, analysis and interpretation, and reflection is required in order to better understand and avoid the trap of subjective conclusions (Guba and Lincoln, 1989, cited in Sohng, 1995; Cornwall and Jewkes, 1995). Validity (the extent to which one is actually measuring what one wishes to measure) and reliability (the extent to which a result or measurement will be the same every time it is measured) remain important considerations. Techniques that were used in this work to improve validity and reliability included the use of standard procedures for data-collection tool development (e.g. in questionnaires), appropriate analysis of different kinds of data (e.g. statistical software for quantitative analysis) and triangulating results obtained through different methods (e.g. use of secondary data sources on health problems). PEAR also considers whether social action arises from the research process and whether the changes made bring about the outcomes observed – so-called catalytic validity (Heron, 1988, cited in Sohng, 1995).

Language was a key issue in interpretation and analysis. French was the working language for this thesis and was a second language both for the lead researcher (a native English speaker) and for the farmers (native Wolof speakers). For data collection, questions were written in French and then translated into Wolof. For the questionnaires, another Wolof speaker from outside the research context translated what he heard into French to assure the words were correctly interpreted. All interventions were first developed in French and then translated into Wolof, and then translated into French for me. The notes taken were a combination of French and English and were eventually transcribed into English. Given the important role that language plays in the interpretation of meanings, the several filters that developed during the research might have influenced the analysis of findings.

RESEARCH FINDINGS

Health and environment impacts linked to urban agriculture

When asked directly what the main environmental problems were in the region in the year 2004, farmers spoke of domestic waste, rotting dead animals, mosquitoes and random defecation in the fields. Other studies have suggested that Dakar's waste-production levels exceed the garbage collection and recycling capacity of the city because most citizens live under precarious conditions (Mbaye et al, 2000). Observations were made of UA's use of substantial amounts of both wastewater and pesticides in the zone.

Farmers reported malaria, parasitic infection, dermatitis and fatigue as the top four illnesses they had had in the past year. In comparison, health district statistics in Pikine cited malaria, dermatitis, parasitic infections, arterial hypertension, diarrhoea and anaemia among the top six diseases for all ages in the district (Pikine District Health Centre, 2005). Interestingly, diarrhoea was not cited among the farmers as a health problem, perhaps because of its greater severity and incidence among children. Under-reporting of diarrhoeal episodes may also occur due to the cultural taboos related to discussing issues related to toilet habits in Senegal. As a result of focus group discussions and findings from initial questionnaires, it was decided that the PEAR process would focus on wastewater and pesticide use in the zone.

Perceptions of wastewater, health and environment risk

In Pikine, farmers use untreated wastewater and what are known as *céanes* to irrigate their fields. Céanes are large, shallow, hand-dug wells up to 3 m deep and 5 m in diameter and are highly saline due to their proximity to the coast and to sea level. Untreated wastewater is often accessed via cracked sewage mains, sometimes broken on purpose by farmers, that carry untreated wastewater. Farmers were observed wading directly into ponds to fill their watering cans, wearing only shorts and a T-shirt. After watering, farmers also touch crops directly, as they weed and plant. In addition to dermal exposure, there is a potential for ingestion as water splashes frequently into the face, and farmers often eat in the fields immediately after watering. Ninety-three per cent of farmers surveyed said that they did not use protective equipment, citing that it impeded work and was too costly. In focus group discussions, farmers mentioned bad odours and nausea as nuisances, but also stated that they became accustomed to them after some time.

When asked to list illnesses potentially related to wastewater, 45 per cent of the farmers said that there were none, almost exactly the same as that in an earlier study – 45.8 per cent of farmers did not consider wastewater as a source of illness (Niang, 2002). Furthermore, 71 per cent of the farmers said that they personally had not suffered any illnesses related to wastewater. Parasitic infections, malaria and dermatitis were, however, among the health effects that were cited for those who had made the connection. A study conducted in 2000 in Pikine, where water sources are mixed (*céane* and raw wastewater), showed that at least 41 per cent of those farmers using *céane* water were infected with intestinal parasites. The most common parasites found were *Ascaris ascaris* (roundworm), *Trichuris trichiura* (whipworm) and *Strongyloides stercoralis* (threadworm) (Niang, 2002). This suggests that many of these farmers may have been infected by parasites without knowing it, partially explaining why they did not initially see the need for education on wastewater effects. The invisibility of contaminants and the lack of connection made between symptoms of potential illnesses and exposure also contribute to the lack of demand for education.

Pesticides, health and environmental risk perceptions and practices

All farmers surveyed used pesticides classified by WHO as Class I, II and III (extremely, highly and moderately hazardous). The top five pesticides used were dimethoate, dicofol, lanate (methomyl), maneb and metaphos (parathion). These were applied liberally and frequently (on average every ten days) during the growing season, on various vegetables, with little regard to the direction of the wind. Farmers were observed to wear very little protective clothing and equipment, exposing arms, hands, legs, feet, face and head to pesticides on a continual basis. Also, they often remained in fields after a treatment despite the presence of strong chemical smells, and were observed smoking, drinking and eating during and after applications. They were also seen weeding and planting after treatments, increasing the potential for exposure to pesticides through contact with air, soil and plants. The half-life of some of these pesticides has been documented at 60–120 days, suggesting the continual presence of these chemicals in the environment (Gueye, 2005).

In addition, according to discussions with farmers, crops are harvested and sold immediately after pesticide application without allowing for the proper waiting period. Women often harvest crops and both children and babies were regularly observed playing in fields. This suggests the potential for pesticide exposures among the larger farm population.

Farmers identified headache and vertigo most frequently as health effects linked to the use of pesticides. This was followed by stomach ache and other digestive and respiratory problems. Twenty-eight per cent (14/50) of the farmers reported a pesticide poisoning in the past year, comparable with another study conducted in the Niayes where 25 per cent of surveyed farmers reported pesticide poisonings (Cisse, 2000). Symptoms related to these poisonings in both studies included vertigo, dizziness, headaches, stomach ache and blurry vision. These are in keeping with the known acute effects of the principal products identified as used just prior to the poisoning, including metaphos, dicofol and methamidophos (International Programme on Chemical Safety, 2006). Such congruence suggests a valid understanding of a cause and effect relationship between pesticide exposure and health effects. Also, it underlies farmers' articulated need for education on the health and environmental effects of pesticides as revealed in focus groups and questionnaires.

The PEAR process

During my work with the farmers, the lead researcher attempted to facilitate a collaborative inquiry process using various participatory techniques in order to identify, analyse and resolve problems. What insights did such a process bring regarding the nature of knowledge and existing power dynamics? What indications of change in understanding of ecosystems occurred? The following sections provide a summary.

The nature of knowledge

At the beginning of the research process, Pikine farmers cited their faith in the chemical companies and their distributors to provide effective chemicals and information on their use. A very small percentage of farmers used alternative methods for pest control. Moreover, those with farming traditions either had lost knowledge on traditional methods of pest control or such methods were not applicable in the high-intensity urban context. According to one farmer, 'observation' is the key method used among experienced farmers. He stated: 'the farmers who observe their plants know best how to manage them'.

However, farmers claimed that they did not know how to use these products appropriately as inadequate information and training had been provided by pesticide companies. Through the process of inquiry, farmers explicitly linked pesticide use and health effects and realized the importance of learning about their safe use. During planning sessions, farmers who acted as spokespersons for the group asked for a train-the-trainer workshop whereby trained professionals would provide initial training on pesticide use for literate farmers. Farmers wanted those who were trained in the scientific method to provide information, which would then be disseminated to other farmers.

When asked whether they believed that there were health effects linked to wastewater, at least 55 per cent of the farmers suggested that there might be some possible health effect linked to its use, such as parasitic infections, dermatitis or malaria. However, 71 per cent said that they personally had had no illness linked to its use. This may be the reason that no farmers asked for education on the safe use of wastewater in either focus groups or questionnaires. Water quality testing carried out by the University of Dakar, however, did confirm the presence of pathogens in the water, and clinical data identified at least 41.8 per cent of the farmers using *céane* water were infected with the same parasites found in the water. Follow-up evaluation showed that although farmers had a greater awareness of health effects linked to wastewater and washed more regularly, their reported illnesses linked to it did not increase. Observation showed very little change between 2005 and 2006 in farmers' watering practices, although questionnaire responses did suggest some change. Several factors influence decisions to protect against environmental health risks, which were discussed with farmers through an exercise examining barriers and solutions. Barriers to behaviour change cited by farmers included: lack of adequate equipment, funds and information, illiteracy, negligence, no experience of any health effects related to exposure, ignorance, fear of loss of productivity and lack of access to and absence of skill in the use of alternatives. This exercise provided the basis for the development of action plans to address these barriers.

Other possible explanations for risk perceptions probably exist. According to Guisse (M. Guisse, 2005, in-depth interview), in Senegal 94 per cent of the population is Muslim and 6 per cent is Christian. Traditional African views of the world are intertwined with these beliefs. For example, humans are considered able to enter into communication with the invisible world of spirits and ancestors through rituals such as dance and singing, in order to heal the sick or accomplish other ends. Following such traditional beliefs, one moves through a cycle of birth, development, death and return to life through reincarnation. Guisse contends that such beliefs influence the way in which people perceive risks.

Pikine farmers expressed the following belief statements to me, during my research: 'One might say that it is a poison that kills... but we say that it is not the poison it is God'; 'Everything that doesn't come from God can do nothing against an African'; 'Most people think it is God who decides. One could be shot by a pistol and not die. You could put a bullet vest on and if God decides you will die, you will die'.

Farmers also referred to the ability of Africans to resist disease and exposure to harmful chemicals and pathogens, particularly in comparison with Europeans. This was illustrated by the story of a farmer who dived into a wastewater pool to fix a pipe and suffered no apparent health effects.

Power relationships

PEAR ultimately recognizes that knowledge is power (Mordock and Krasny, 2001). One might suggest that the pesticide companies play a role in controlling access to information. For example, container labels often have insufficient or incorrect information on hazard class and use, perhaps in order to encourage their sale. According to the Sahelian Pesticide Committee of the Comité Permanent Inter-états de Lutte Contre la Secheresse (CILSS), the hazardous pesticides used by farmers are not authorized for use in Senegal. But rules about pesticides are unclear and are currently changing, making it difficult for government officials to enforce any regulations covering pesticide imports, sales and use (Williamson, 2003).

National legislation on pesticides may not even help, however, because, although technically approval will be required, in reality existing pesticides are cheap and readily available. The possibility of increased black market activity to access pesticides also exists because they come over the border from neighbouring countries that have less-strict controls (H. Diouf, 2007; PAN-Africa, personal communication).

Despite the known toxic effects of many pesticides, global corporate economic interests are still supplying these products on the international market and selling to local formulators (Kishi and Ladou, 2001). In Pikine, pesticides are sold by Senchim and Traore, Senegalese agro-chemical formulators. Very close relations exist between farmers and officials of Senchim; for example, Mr Sissokho, a farmer and also a municipal politician who was one of the driving forces of the PEAR process, was in contact socially with Senchim officials everyday. Although no evidence suggests that this relationship influenced farmers to use more toxic products, potential could have existed for this or vice versa for farmers to influence the formulation of less-toxic products.

According to Guisse (M. Guisse, 2005, in-depth interview), Senegal is primarily a nonindustrial culture where the sense of responsibility to the industrial environment is limited. According to Guisse, if an accident occurs from pesticide exposure, no collective management body (the state, district medical officer, ministry of health, pesticide distributor) is mandated to intervene. In fact, farmers are alone in the world and are victims (M. Guisse, 2005, in-depth interview) unless they organize with others. The GIE Provania has attempted to address these power issues. The secretary, for example, has a diploma in horticulture and is a municipal politician. Because he is the group's driving force in regard to partnership and project development, he wields the most political influence. However, because he is the quickest to respond to questions about process, content and what action to select to address problems, others of the group do not always have the time to provide their input.

Participation

Extension workers commented that the GIE Provania had brought together all the farmer associations in the area on the issue of environment and health. Participation of farmers was facilitated by close collaboration with the executive of GIE Provania, who were local community leaders. All had at minimum a secondary education, which allowed them to access information easily and communicate with me in French, as well as act as translators and facilitators during the PEAR process. These leaders used the internet to access websites which were pertinent to the issues being addressed, including information on pesticide toxicity.

Having the GIE Provania executive team, who are farmers, administer questionnaires to other farmers may have produced some biases because translation of Wolof terms might not have been consistent. In addition, farmers who were trainers also administered evaluation questionnaires, which may have skewed responses through the prompting of respondents by the interviewer. Because of my inability to speak Wolof, I was unable to speak directly to the most marginalized, illiterate farmers. Similarly, my involvement was exclusively with men, and I did not work with women, despite their important role in marketing produce and their potential exposures to pesticides and wastewater. Working with this population might have been possible in small groups and using a translator. The GIE Provania executives, however, provided a legitimate brokering role, enabling messages to spread more rapidly in order for action to take place.

Mapping changes

The first time farmers constructed a map, they focused on physical features of the site: a golf course, a body of water, fields, an experimental water filtration basin, typha plants, mosquitoes and the main road through the fields. Farmers explained that bodies of water where typha grows become stagnant, killing the fish that would otherwise eat mosquitoes. They also said that the water found between the golf course and the fields used to be fresh but has become increasingly saline.

One year later, the farmers made three separate drawings, which they used to describe several relationships between health and the environment. Some messages that had been discussed in the past were apparent, such as the danger associated with inhalation exposure during spraying; avoiding spraying under a strong sun because skin pores are more open and absorb pesticides more easily; and avoiding walking in fields that have been treated. In one drawing, three different plots with *céanes* were depicted with a creek and fish. A farmer is shown spraying fields with metaphos and pesticide run-off because of an incline into the water. As fishing is a common activity in the zone, farmers were concerned about killing or contaminating fish that would later be consumed. This demonstrated their understanding of pesticide mobility into the food chain. Therefore, they understood there were dangers beyond those which might only affect their own persons, indicating that they were extrapolating risks to the surrounding community residents and children (Figures 10.6 and 10.7).

The following maps drawn by farmers illustrate more fully the dynamic interaction between human activity, agricultural practices and impacts on environment and community health.

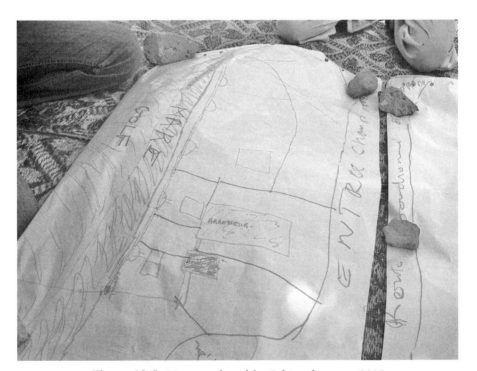

Figure 10.5 *Map produced by Pikine farmers, 2005*

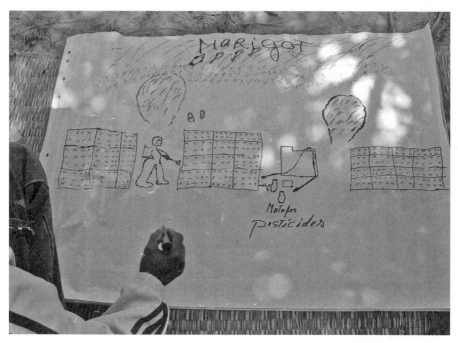

Figure 10.6 *Map 1 produced by farmers, 2006*

Figure 10.7 *Map 2 produced by farmers, 2006*

Changes as a result of the PEAR process

The ability to use interactive techniques to convey messages on environment and health has developed significantly among farmer trainers. In fact, during training sessions, participating farmers demanded that such techniques be used. They remarked that systematic questioning to facilitate issue identification and problem solving provided them with new skills to organize their thoughts and reflect on issues. The use of existing skill sets, such as drawing illustrations, has also reinforced group capacity.[2]

The link between environment and health and agricultural practices has become an important theme for GIE Provania. It has been incorporated into their daily activities and integrated into their work in the intra-urban wetlands (Niayes) zone. A significant reduction in waste pesticide containers has been noted in the fields, which suggests the possibility that fewer pesticides are being used and that improved disposal methods have been adopted. As a result of discussions between the researcher and GIE Provania, a Senegalese pesticide formulator named Senchim has entered into discussion with the farmers so as to develop a waste-collection system for empty pesticide containers.

As a result of analytical exercises to identify barriers and solutions in order to address environmental health issues, the GIE has developed action plans, including greater development of techniques to avoid pesticide exposure, such as IPM, more prudent crop choices, use of personal protective equipment and first aid. Farmers are currently seeking funding for these activities in partnership with such organizations as ANCAR, ENDA Rup, IFAN, ISRA and other GIEs.

CONCLUSIONS

Data on environmental and health risk perceptions suggest that farmers are aware of pesticides' health effects, as illustrated by the congruence of reported symptoms with known health effects. This was not so evident for wastewater. The use of participatory techniques to transmit information on health and the environment and strategize for action is a new approach in the Niayes, which has been adopted by farmers.

Farmers' perceptions on the relationship between malaria and UA figure prominently among the themes for additional research. Further research is needed in order to investigate the contribution of UA and peri-urban agriculture (UPA) to anopheles densities (Klinkenberg et al, 2005). More emphasis should be given to learning regional languages and verifying translations to ensure access to the most marginalized groups and greater validity of research findings. In addition, more time in the field is required so that research agendas with local researchers could be better coordinated. Such time could also improve the interpretation of findings. Furthermore, more PEAR studies should be implemented to assess the contributions of different approaches, as well as of tools for joint learning and action in UA settings.

ACKNOWLEDGEMENTS

I would like to thank Dr Donald Cole for his support throughout this research process, as well as Dr Seydou Niang for insights and logistical support in the field. I would also like to thank Oumar Sissokho, Pape Diagne and Amadou Dia, executive members of GIE Provania, whose dynamism and desire to improve the lives of fellow farmers made this research possible. Funding for this research was made available by the International Development Research Centre's (IDRC) AGROPOLIS award programme.

NOTES

1 Nita Chaudhuri, Ph.D. (candidate), University of Toronto, 16 allée du vieux jardin, 78290 Croissy sur Seine, France, email: nita.chaudhuri@utoronto.ca
2 Other doctoral students have also been influenced by this process. In response to the fatigue she felt farmers expressed about researchers who collect information but provide nothing in return, a soil science student from the University of Lausanne incorporated an interactive education component into her research on land degradation.

REFERENCES

Alkorta, I., Albizu, I., Amezaga, I., Onaindia, M., Buchner, V. and Garbisu, C. (2004) 'Climbing a ladder: A step by step approach to understanding the concept of agroecosystem health', *Reviews in Environmental Health*, vol 19, no 2

Cisse, I. (2000) 'Utilisation des pesticides dans le système de production horticole dans la zone des Niayes: les produits et leurs impacts sur la nappe phréatique', Thèse de 3^{ieme} cycle Géographie, UCAD, Dakar, Senegal

Cornwall, A. and Jewkes, R. (1995) 'What is participatory research?', *Social Science & Medicine*, vol 41, no 12, pp1667–1676

Editions, J. A. (2000) *Les Atlas de l'Afrique, Atlas du Sénégal*, 5th edition, Editions J. A., Paris, France

Gueye, A. (2005) 'Assessment of environmental impacts due to the reuse of untreated wastewater in periurban agriculture of Dakar, Senegal', final report for the Young Scientists grant, KFPE, Université de Lausanne, Lausanne, Switzerland

International Development Research Centre (IDRC) (2006) *Shaping Liveable Cities – Stories of Progress around the World*, IDRC, Ottawa, ON, Canada

International Programme on Chemical Safety (2007) International Chemical Safety Cards, WHO, Geneva

Kishi, M. and Ladou, J. (2001) 'International pesticide use', *International Journal of Occupational and Environmental Health*, vol 7, no 4, pp259–265

Klinkenberg, E., McCall, P. J., Hastings, I. M., Wilson, M. D., Amerasinghe, F. P. and Donnelly, M. J. (2005) 'Malaria and irrigated crops, Accra, Ghana', *Emerging Infectious Diseases*, vol 11, no 8, pp1290–1293

Lebel, J. (2003) *Health: An Ecosystem Approach*, International Development Research Centre, Ottawa, ON, Canada

Mbaye, A. and Moustier, P. (2000) 'Market oriented urban agricultural production in Dakar', City Case Study, Dakar

Mooney, H. A., Lubchenco, J., Dirzo, R. and Sala, O. E. (1994) 'Biodiversity and ecosystem functioning: ecosystem analyses', in Heywood, V. H. (ed) *Global Biodiversity Assessment*, Cambridge University Press, Cambridge, UK

Mordock, K. and Krasny, M. E. (2001) 'Participatory action research: A theoretical and practical framework for EE', *Journal of Environmental Education*, vol 32, no 3, pp15–20

Niang, S. (2002) 'Maitrise des risques dans la valorisation des eaux usées en agriculture urbaine', in Akinbamijo, O. O., Fall, S. T. and Smith O. B. (eds) *Advances in Crop-Livestock Integration in West African Cities*, International Trypanotolerance Centre, ISRA, IDRC, Ottawa, Canada

Niang, S. and Gaye, M. (2004) 'L'epuration extensive des eaux usées pour leur reutilisation dans l'agriculture urbaine: des technologies appropriées en zone sahélienne pour la lutte contre la pauvreté', Phase II: Rapport Scientifique Provisoire 3, IFAN and ENDA (2006) Dakar, Senegal

Pikine District Health Centre (2005) Pikine District Health Statistics

Sohng, S. S. L. (1995) 'Participatory research and community organizing', www.cdra.org.za/articles/Participatory%20Research%20And%20Community%20Organizing%20by%20Sung%20Sil%20Lee%20Sohng.dec (accessed December 2004)

UNDP (2005) Human development report, United Nations, New York, NY

United Nations Centre on Human Settlements (UNCHS) (2001) *Cities in a Globalizing World – Global Report on Human Settlements*, Earthscan, London, UK

Williamson, S. (2003) *The Dependency Syndrome: Pesticide Use by African Smallholders. Pesticides, Poverty and Livelihoods*, Pesticide Action Network (PAN-UK), London, UK

11

Complex Ecologies and City Spaces: Social–Ecological Networks of Urban Agriculture

Laura J. Shillington[1]

INTRODUCTION

Urban agriculture is about more than just growing food in cities; it has to do fundamentally with human–environment interactions and relations in urban areas. How urban residents relate to and interact with the natural environment in cities shapes how UA materializes, and these relations differ between cities and between residents of the same city. Yet this aspect of UA has rarely been explored despite the fact that agriculture in cities presents an excellent situation for examining human–environment relations. The broader issue of urban human–environment relations has received little attention primarily due to the way in which the urban and rural are generally understood: as separate spaces at opposite sides of the social (human)/nature divide. This separation arises out of dominant Western conceptualizations that have structured the world around binaries (Demerrit, 2002). Accordingly, the urban is considered a predominantly human (social) space outside of nature, whereas the rural is the space of 'nature' (Wolch et al, 2001; Braun, 2005; Swyngedouw, 2005). While there is ample research on human–environment relations in rural areas, especially in political, ecological and earlier peasant studies literature, there is a lack of similar analyses on human–environment relations in cities. Urban agricultural research has, for the most part, continued to maintain the binary of urban and rural. For instance, a main concern has been to ask what is urban about *urban* agriculture, thus equating agriculture as a fundamentally rural activity. Yet such a question is rooted in the binary of urban–rural and in a lack of understanding about human–'natural' environment relations in cities. Indeed, this is visible in many urban agricultural studies which link agriculture taking place in cities to that in rural areas, and suggests that this link to rural 'natural' areas is significant to the presence and continuance of UA (Frayne, 2004). While such a connection might be critical for urban agricultural practices in many cities, there exist more diverse relations between human and environment in cities that play an important role in UA.

Drawing on the current situation in Managua, Nicaragua, this research examines existing relations between human and environment at the household scale. It describes how humans interact with the environment in the establishment and management of gardens and food production in urban areas. Managua provides an especially interesting case study because it represents in many ways a city that defies dominant notions of 'the urban' (Wall, 1996). Managua has often been referred to as un-urban, primarily because it was never fully rebuilt after the devastating earthquake in 1972, and as a consequence the contemporary urban form does not resemble a typical 'urban' centre (McGuire, 1991; Wall, 1996; Rodgers, 2004). Most noticeably, Managua's built landscape is unusually horizontal with very few buildings exceeding five storeys. A large make-up of this landscape are houses, which are the most common type of dwelling in the city. In fact, detached houses account for 97 per cent of the dwelling types in the city (INEC, 2006).[2] The majority of these houses have yards large enough to grow trees and plants, and most have at least one tree. As a result, Managua's cityscape is remarkably green. This research looked at the small-scale agriculture and agro-forestry within home patios (yards) in San Augusto,[3] a marginalized barrio of Managua. It examined the relations through which these forms of UA and gardening are produced and what role they play in the development of sustainable urban spaces. Additionally, the study examined whether the human–environment relations used and created in UA assist in generating community-based actions around social and environmental change in the barrio. The concept of 'networks' is used to explain the interactions between humans and the environment, which are referred to in this chapter as socio-ecological networks.

OBJECTIVES, HYPOTHESES AND CONCEPTS

The objectives of this research were: (i) to better understand the human–environment relations involved in UA by describing the socio-ecological networks; (ii) to determine the importance of these networks and relations in producing sustainable urban spaces; and (iii) how these networks and relations impact on the livelihood strategies in poor barrios of Managua. The study attempted to answer the two main questions. First, what socio-ecological networks exist in UA and how important are these in creating liveable home and community spaces? Second, do these socio-ecological networks in UA play a role in development of social organizations or movements?

Based on the assumption that individuals create distinct socio-ecological networks in household-scale UA activities, the hypothesis of this study was that differences in household urban agricultural practices and forms do not necessarily hinder the way that communities attend to social and environmental challenges.

In this study, human–environment relations are seen as networks comprising of interactions between people, plants and other living creatures. Gender and development studies have shown that individuals and households depend on informal social networks for their livelihood strategies (Moser, 1993; Peake and Trotz, 1999; Silvey and Elmhirst, 2003). These kin and friendship networks enable individuals to acquire and share resources, and to assist with their livelihood strategies (Peake and Trotz, 1999).[4] Indeed, it has been noted that these kin and friendship networks have engendered collective efforts to empower women in their everyday lives and in politics (Moser, 1993; Kabeer, 1994; Agrawal, 2000).

Recently, work in feminist geography and political ecology has argued that such networks are not only social, but that they are also connected to particular ecologies such as plants, trees and other natural resources. As Ro (2001, p78) explains, socio-ecological networks are 'patterns and processes of habit-forming connections between people (individuals and groups), other beings, physical surroundings and artefacts'. This understanding of networks emphasizes the complex ways that people deal with social and environmental change. That humans rely on both social and ecological elements to deal with livelihood struggles is not a new recognition, but this reliance goes beyond food and economics (Rocheleau, 2001; Wolch et al, 2001; Page, 2002).

METHODOLOGY

Data collection

Research methods are aimed at gathering information on relations and interactions between individuals and plants in the households, and their influence on the development of sustainable UA in San Augusto, a poor, residential area in Managua, Nicaragua. San Augusto is a barrio that was settled informally during the mid-1990s. It is located well inside the legal city limits, but is at the southern edge of urban development in the city (Figure 11.1). This barrio is among the poorest in Managua and remains without most urban services, with limited infrastructure.

The research involved two different methods. Primary data was obtained through interviews with relevant government departments and non-governmental organizations (NGOs), community-based interviews, participatory mapping and focus groups. Secondary data was gathered by a review of literature and archival research. The collection of secondary data involved examination of different current and historical policies and projects. However, due to difficulties experienced in accessing research materials and documents from the archives, official government and non-governmental interviews became the main source of information. At the community level, two NGOs, the Fundación Nicaragüense Pro-Desarrollo Comunitario Integral (FUNDECI) and

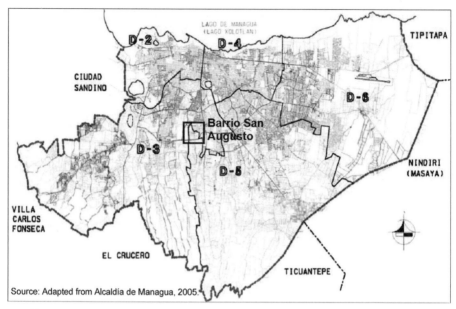

Figure 11.1 *Map of Managua showing location of Barrio San Augusto*

the San Augusto Community Association, Asociación Comunitario San Augusto (ACSA), assisted in mobilizing community members. FUNDECI also participated in data collection, focus group discussions, interviews and in participatory mapping.

Semi-structured interviews and participatory mapping were the main methods through which primary data at the community level were collected. In San Augusto, there were a total of 25 community members (20 women and five men) who participated in the interviews and mapping process. The sample population was obtained through two means: first, by asking the participants of an urban agricultural workshop organized by FUNDECI; and second, through invitations sent (in person) to households not participating in the workshop. While the research interviews and mapping were directed equally at men and women in households, the majority of participants were women.

There are several main reasons why more women participated. First, FUNDECI's urban agricultural workshops were aimed at encouraging small-scale food production in home patios, an activity traditionally undertaken by women in Nicaragua (Lok, 1998; Méndez et al, 2001). Second, patios in these urban homes are more than just gardens or yards. Instead, they are 'rooms' of the house – that is, they contain the kitchen, laundry area, washroom and living room where women's domestic tasks take place. Therefore, the patio is viewed as a female space. In Nicaragua, women continue to be responsible for the majority of domestic chores (Espinosa, 2004).[5] As such, women are the primary caretakers of that space, even when men also carry out activities in the patio. Men, for example, may help by trimming trees, repairing fences and other structures, doing animal husbandry, repairing electronics and managing home-based stores (FUNDECI, 2004). Consequently, when the invitations were given to households to participate in the research, the men would pass on the invitation to the women and it was usually the women who agreed to participate.

The purpose of the mapping and interviews was to document the socio-ecological networks used in UA and to make inventory histories of the patios. The mapping process was carried out in a participatory manner where participants identified the various plants and trees in their yards and mapped them by sketching these spaces. The action of mapping was used to provide a detailed chronology of the plants and trees, where at each plant or tree mapped the participants would relay how the plants were obtained, their purpose and use, as well as who is involved in their management. The interviews complemented mapping by providing more detailed socio-economic information of the household.

Focus groups were utilized to gather data on collective interests. A total of 50 participants were involved in three separate focus groups.[6] Two focus groups were conducted at the beginning of one of FUNDECI's *huertos familiares* (family gardening) workshop.[7] Topics discussed included the importance of plants and trees in patios, the different ways in which people produce and maintain their patios, and opinions regarding growing food in patios, communities and cities in general. At the end of the project, the group was involved in evaluating the FUNDECI project and this study.

Data analysis

All of the interviews, mapping and focus groups were tape-recorded; later these were transcribed and analysed. Large visual data maps, which consisted of large pieces of brown paper with charts outlining different categories, were based on data (e.g. economic, social) from interviews. Information from the participatory mapping exercises was analysed in a similar way, creating different categories. The categories that emerged from the mapping process formed the basis for describing socio-ecological networks (Table 11.1), which were then used to further analyse the social and economic information from the interviews. All data was disaggregated by gender as well as by plant type.

RESEARCH FINDINGS

Diversity and difference in patio plants

The patios in San Augusto have a large diversity of plants and trees. During the mapping process, a total of 713 different plants were identified. While this is not a complete inventory of all the plants and trees, the mapping provided a good representation of what people's patios contain. Women, who constituted the majority of the participants, mapped over five times more plants than

Table 11.1 *Categories of networks from mapping process*

Main category	Sub-category	Explanation
Purchased	Seedling or seeds	The seeds or small plants (seedlings) are purchased (monetary exchange) at a market, street stall, *pulpería* (corner store) or other place
	Fruit	The fruit or vegetable is purchased through monetary exchange and then the seeds from that fruit are purposively planted
Given	Plant (seedling), seeds or fruit given by family member	The plant, seed or fruit is given by a family member. If fruit given, then seeds from that fruit are purposively planted
	Plant (seedling), seeds or fruit given by family friend	The plant, seed or fruit is given by a friend. If fruit given, then seeds from that fruit are purposively planted
	Plant (seedling), seeds or fruit given by neighbour	The plant, seed or fruit is given by a neighbour. If fruit given, then seeds from that fruit are purposively planted
	Plant (seedling), seeds or fruit given at workplace (by employer, co-worker, etc.)	The plant, seed or fruit is given by someone at place of work. If fruit given, then seeds from that fruit are purposively planted
	Plant (seedling), seed or fruit is given by a stranger	The plant, seed or fruit is given by a stranger. If fruit given, then seeds from that fruit are purposively plants
Found	Seedling taken from a street or other public place	A seedling or cutting from plant is taken from the street, park, main road, etc. (any public place)
	Seedling taken from stranger's patio	A seedling or cutting is taken from a stranger's patio (without them knowing)
Other	Plant already there	The plant was there when household was started
	Grew naturally	Either the plant grew from seeds in household rubbish or by other means whereby no one in household purposively planted it

men. Table 11.2 presents the number of plants identified and mapped by type. The participants categorized their plants and trees according to their primary use: fruit, medicinal or spice, vegetable, ornamental and shade trees. This study utilized these categorizations as opposed to taxonomic classification.

Of the total number of plants identified, fruit trees and ornamentals accounted for 34 per cent and 32 per cent respectively. Indeed, this is evident in the landscape of the neighbourhood, which has a well-established canopy

Table 11.2 *Total number of plants by category in patios*

		Total number of plants	Fruit trees	Non-fruit (shade) trees	Vegetable plants	Medicinal/ spice plants	Ornamental plants
Men	5	119	51 (43%)	12 (10%)	0 (0%)	21 (18%)	35 (29%)
Women	20	594	189 (32%)	34 (7%)	48 (8%)	129 (22%)	194 (33%)
Total	25	713	240 (34%)	46 (6%)	48 (7%)	150 (21%)	229 (32%)

of avocado and mango trees, the most commonly mapped trees. While fruit trees and ornamentals constituted the majority of plants in all patios, men tended to map fruit trees and women ornamentals. Out of the plants that men mapped, 43 per cent were fruit trees and 29 per cent ornamentals. Fruit trees accounted for 34 per cent of the plants mapped by women, while ornamentals were 32 per cent. This difference was not as significant as that between shade trees (non-fruit) and vegetables. Men identified more non-fruit trees than women. However, at times, women ignored the non-fruit trees in the patio while, in some cases, men ignored certain fruit trees. In some houses where women participated in the mapping, non-fruit trees were rare and vegetable plants more common. In contrast, where men carried out the mapping, some of the patios had both fruit and non-fruit trees but seldom had any vegetables. None of the men mapped vegetable plants, while 7 per cent of plants women mapped were vegetables.[8]

Although the total number of plants identified by all participants was 713, that included only 130 different plant species (Table 11.3). This is because many patios had multiple plants of the same species.

For example, in one patio (Figure 11.2), the participant mapped 14 fruit trees in total, but only six different species. There were two avocado trees, a mango, two guava, two lime trees, six banana/plantain plants and a coconut tree. Within the banana/plantain plants, there were three different varieties. This was also true of the avocado, mango and lime trees. Each of these species and varieties had, in many cases, a different use due to cultural factors. The stories that participants related about different species also reflected distinct relations and interactions with individuals, environment and community. This was true especially with ornamentals; because they reproduce easily they are often shared among the members of different networks.

On average, women identified 22 different species in their patios while men mapped 13. This difference was a result of the large number of ornamental plants that women mapped; in their patios ornamentals comprised 54 per cent of the different species. Men tended to map trees and the species-diversity in what they mapped comprised 40 per cent fruit and 18 per cent non-fruit trees. By comparison, species diversity in women's maps consisted of 22 per cent fruit

Table 11.3 *Species diversity identified by plant type*

Plant type	Total number of different species identified within plant types	Number species identified within plant type (% of total species identified)	
		Men	Women
Fruit trees	31	16	27
		(40)	(22)
Non-fruit trees	13	7	10
		(18)	(8)
Vegetable plants	9	0	9
		(0)	(7)
Medicinal/spice	11	6	11
		(15)	(9)
Ornamental plants	66	11	66
		(28)	(54)
Total number of species identified	130	40	123
		(31)	(95)

trees and only 8 per cent non-fruit trees. Overall, women identified 123 out of the 130 species identified, being 95 per cent of the total species diversity of the patios.

While there was a gender difference in the plants and trees identified, the ecological composition of the patios was similar. The majority of the patios were dominated by a canopy of large fruit trees, interspersed with the occasional non-fruit tree, with an understory covered with ornamental plants. The gender difference, therefore, was not in the composition of patios but in the identification of plants and trees. What the participants identified illustrates gendered differences in human–environment relations.

The results suggest that women consider ornamental plants and fruit trees to be the most important plants in the patios, both in number and species diversity. Their benefits for the household were numerous, despite the fact that neither provide any significant income. Women commented that having plants creates an aesthetically pleasing space, and a more homely and comfortable environment. Both women and men explained that plants and trees are also important in the creation of private spaces, as most of the everyday activities (e.g. cooking, bathing) in San Augusto take place outside on patios. In addition, they provide a way to conceal the haphazard construction of the houses.[9] Therefore, ornamental plants offer a quick way to produce liveable home spaces, as they are easy to grow and propagate rapidly. Moreover, ornamentals are, in most cases, the only plants that flourish under the shaded canopy of fruit trees that dominates patios because food crops are difficult to grow under such conditions. For this reason, very few vegetables were present in patios.

Fruit trees were another important species in patios for both women and men, and were planted for cultural, environmental and livelihood reasons.

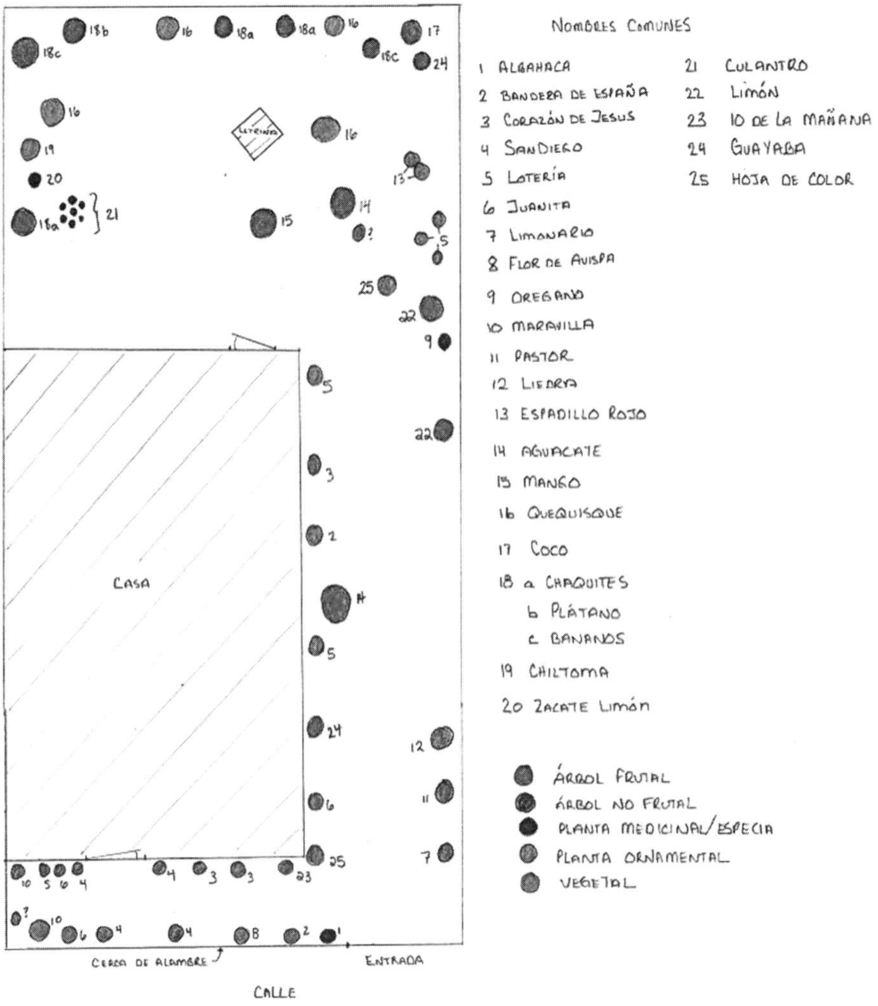
Nombres Comunes
1 Albahaca
2 Bandera de España
3 Corazón de Jesus
4 San Diego
5 Lotería
6 Juanita
7 Limonario
8 Flor de Avispa
9 Oregano
10 Maravilla
11 Pastor
12 Liedra
13 Espadillo Rojo
14 Aguacate
15 Mango
16 Quequisque
17 Coco
18 a Chaquites
b Plátano
c Bananos
19 Chiltoma
20 Zacate Limón
21 Culantro
22 Limón
23 10 de la Mañana
24 Guayaba
25 Hoja de Color
Árbol frutal
Árbol no frutal
Planta medicinal/especia
Planta ornamental
Vegetal
Letrina
Casa
Cerca de alambre
Entrada
Calle

Figure 11.2 *Map of patio in San Augusto*

The most common fruit trees were mango and avocado, followed by papaya, coconut and banana/plantain. Along with these, almost all patios had a different combination of fruit trees, such as lime, orange, guava, sapote, *mamón* (commonly known as 'honeyberry' in English), jocote (a tropical plum) and *marañon* (cashew). The majority of these trees are relatively large, producing shade in patios and houses. This is critical in Managua, where daytime temperatures average 32°C year round. Thus, large fruit trees such as mangoes and avocadoes protect the houses and patios from harsh sun and heat, keeping them cool.

In fact, respondents noted a main attribute of San Augusto as being the cool temperatures compared with other areas of Managua. Yet, although fruit trees were commonplace, more than 95 per cent of respondents said most of the fruit was used only for household consumption and any surplus was given to neighbours. Very few respondents sold fruit from their patios; some respondents explained that they only sold to strangers.

Non-fruit trees and medicinal plants and herbs are also part of the ecology of patios in San Augusto, although they were not as prominent as fruit trees and ornamentals. In some cases, medicinal plants and herbs were difficult to grow because of the shade generated by large fruit trees. Most of the women did not consider this a problem, however, because they found small sunny areas at the margins of their patio to grow commonly used medicinal plants and herbs such as *albahaca* (basil), *culantro* (variety of coriander), chillies[10] and *hierba buena* (mint). Such plants contributed to household food consumption and health care. For example, basil is used to treat earache.

The diverse composition of plants and trees in patios has produced particular human–environment relations, which simultaneously were a result of socio-ecological networks of household members, neighbours, family and other environments. There were clear differences between the patios that men and women mapped; women on average mapped and identified more plants and greater diversity of species, suggesting that they pay more attention to the patios. Indeed, patios tended to reflect women's desires and needs. That women were the main participants in the urban agricultural workshops indicates the patios were managed through women's socio-ecological networks.

Network assemblages at work: Connections and relations

The socio-ecological networks of participants were identified from the stories that they told about the plants and trees in their patios. The categories outlined in Table 11.4 illustrate different relations and interactions that participants utilized in creating their patios. There were four main ways in which participants obtained the plants and trees in their patios: purchased, given, found or other (grew spontaneously). Within these main categories, there were several more specific ways. Table 11.4 presents the proportion in each, disaggregated by gender and plant type.

These categories illustrate the most frequently used and most important socio-ecological networks for households. The majority of fruit trees in patios (37 per cent) were obtained through the purchasing of fruit, usually at Managua's largest market, Mercado Oriental. The seed is planted in the patio after the fruit is consumed. In contrast, ornamentals were rarely purchased; 8 per cent had been bought, while 65 per cent had been given by friends, family, neighbours or strangers. Women obtained and exchanged all types of plants and trees through their social networks far more frequently than men. For example, women acquired 25 per cent of ornamentals and 13 per cent of fruit trees from neighbours, whereas men obtained none of the fruit trees and only 6 per cent of ornamentals from neighbours. Common to both women and

Table 11.4 *Distribution of network categories in San Augusto by plant type*

Main category	Sub-category	Fruit trees (%)			Non-fruit trees (%)			Medicinal/ spice plants (%)			Ornamental plants (%)		
		F	M	T	F	M	T	F	M	T	F	M	T
Purchased	Seedling or seeds	2	13	5	0	0	0	17	0	16	7	6	7
	Fruit	35	40	37	0	0	0	20	0	18	0	6	1
Given	Plant (seedling), seeds or fruit given by family member	6	17	9	0	13	6	12	10	18	12	18	13
	Plant (seedling), seeds or fruit given by family friend	14	20	15	10	13	11	22	0	20	16	29	18
	Plant (seedling), seeds or fruit given by neighbour	13	0	10	20	13	17	15	0	14	25	6	21
	Plant (seedling), seeds or fruit given at workplace (employer, co-worker)	0	0	0	0	0	0	5	0	5	1	0	1
	Plant (seedling), seed or fruit is given by a stranger	3	0	2	0	0	0	0	0	0	11	18	12
Found	Seedling taken off street/ other public place	8	3	7	20	63	39	2	0	2	14	12	13
	Seedling taken from stranger's patio	2	0	2	0	0	0	0	0	0	7	0	6
Other	Plant already there	9	3	7	0	0	0	0	0	0	5	6	5
	Grew naturally	8	3	7	50	0	28	7	0	7	1	0	1

men was their procurement of tree seedlings, cuttings or tree branches from sides of streets, in the main road that runs through San Augusto, in public parks and from private properties.

The most prevalent socio-ecological networks were those of family, friends and neighbours. This corresponds with studies in gender and development emphasizing the importance of informal social networks in livelihood strategies (Moser, 1993; Peake and Trotz, 1999). Women were the main participants in UA workshops and the interviews and mapping phases, implying for this study that establishment and maintenance of patios in San Augusto is primarily carried out by women. However, as much as women's socio-ecological networks comprised family, friends and neighbours (primarily within Managua), there was no collective initiative to create community gardens. This was mentioned in several focus groups, and most women and men responded that no-one in San Augusto had proposed such an idea. Moreover, even when the neighbourhood was being established, no area was set aside for communal use.

Urban agriculture in Managua: Gender, private spaces and community efforts

The results of this study indicate that food production in San Augusto is primarily a household activity and is limited to the cultivation of fruit trees and shade-tolerant crops such as herbs and chillies.[11] There was a limited amount of community-based agricultural practices, most of which were informal. At a formal level, the Office of Women for the City of Managua, in collaboration with the United Nations Food and Agricultural Organization (FAO), was involved in several school-garden pilot projects to produce organic vegetables for school kitchens. Other school-garden projects were also being implemented by the Ministry of Education, Culture and Sports (El Ministerio de Educación, Cultura y Deportes, MECD) and the Emergency Social Investment Fund (Fondo de Inversión Social de Emergencia, FISE).

Most efforts to promote UA focus on the scale of the household. FISE, jointly with FAO and the Institute of Aquaculture Technology (Instituto de Tecnología Agropecuaria, INTA), distributed 24,000 tomato seedlings to 300 families in nine barrios, together with tools and fertilizer (Guerrero, 2005; R. Ramirez, FISE representative, 12 November 2005, personal interview). The objective was to encourage households to grow tomatoes for household consumption. If this is successful, the project may expand to incorporate other food crops. Another household project was FUNDECI's *huertos familiares*, which also sought to encourage the establishment of home gardens to assist in enhancing food security and nutrition. However, at the end of FUNDECI's project, participants stated that while they were interested in growing food crops in their individual patios, they did not want to change the composition of trees and ornamentals. Growing food crops that FUNDECI and others promoted would have meant reducing the shade in their patios, consequently diminishing the aforementioned benefits. Thus, many participants commented that having a community-driven effort would have been more successful and beneficial to them.

CONCLUSION AND RECOMMENDATIONS

Gender differences are critical to how UA is carried out and the benefits it produces (Anosike, 2004; Hovorka, 2005, 2006). This study corroborates these authors' findings, showing that men and women have distinct relationships with their environment in the plants they identified and their socio-ecological networks. These varying relationships influence the composition of patios and the extent to which men and women interact with their immediate and surrounding environments. In San Augusto, urban agro-forestry in household patios was the main UA system, and was produced through a diversity of socio-ecological networks (friends, family, strangers, private and public space) and exchanges (monetary, planting materials and harvest exchanges). These networks play an important role in the production of patio gardens, and the ecological composition of patios are critical in creating more liveable homes. The benefits of patios are not limited to food production; the everyday human–environment relations encompassed not only food and economics, but equally the need of comfort and aesthetics. Indeed, the comfort and aesthetics of patio ecologies extends beyond the household to the neighbourhood. As virtually all participants commented, San Augusto's 'cool' environment is unique in Managua, which speaks to the indirect ecological benefits derived from the 'urban forests' of the fruit-tree dominated landscape of the neighbourhood. This is especially true of Managua, where the majority of urban houses have unusually large yards, and where human relations within the environment of these patios are as diverse as the range of urban ecosystems.

ACKNOWLEDGEMENTS

I would like to acknowledge the support of all the staff at FUNDECI in Nicaragua for assisting with this research. Also, I want to thank all the participants in San Augusto. This research was carried out with financial support from Canada's International Development Research Centre (IDRC) through an AGROPOLIS award for urban agricultural research and a Social Science and Humanities Research Council (SSHRC) doctoral fellowship. The research fulfilled part of the requirements of a Ph.D. degree from the Department of Geography, York University, Toronto, Canada.

NOTES

1   Laura J. Shillington, Ph.D., Department of Geography, York University, Calif., USA, email: lshillin@yorku.ca
2   According to the 2005 census, approximately 97 per cent of Managua's population live in houses, 0.12 per cent in apartments, 0.6 per cent in collective housing (e.g. in boarding houses), while 2.68 per cent have no shelter or are living in improvised shelters (INEC, 2006).
3   The community name and the participant names have been changed to ensure anonymity.
4   Development theory has sometimes referred to this as social capital. However, social capital connotes something different from the networks to which I refer. For a discussion on social capital see Das (2004).
5   According to the most recent 2006 national census (carried out by the Instituto Nacional de Estadísticos y Censos), 30.2 per cent of women engage exclusively in productive labour (classified in the census as ama de casas or home caretakers) compared with 1.72 per cent of men. The percentage of women and men engaged in productive labour is respectively 38.4 and 58.7 per cent (INEC, 2006). Note that these statistics include all males and females older than the age of ten years. Espinosa (2004) points out that while many women engage in productive work, they also still tend to be responsible for the majority of domestic tasks in the household.
6   Each focus group was partially tape-recorded and detailed field notes were taken.

7   Not everyone in the workshops was able or willing to participate in interviews and mappings.
8   Any vegetable plants/gardens established with seeds from FUNDECI's project were not mapped, because they were part of a current process.
9   Most houses are built using a variety of materials, chiefly scrap pieces of metal, plastic, concrete blocks, wood and, in extreme cases, cardboard (much of which is scrounged, however, not bought). A form of construction is concrete block foundations and wooden or corrugated metal walls. This type of construction is referred to locally as *minifalda*, which literally translates to 'miniskirt'.
10   Participants classified spicy chillies as a herb and not a vegetable, most likely because it was used as flavouring in cooking.
11   Although chilli plants require direct sunlight, as with many varieties belonging to the nightshade family, too much sun can reduce the production of and damage fruit. Thus, as stated above, many households find small areas in patios that receive a limited amount of direct sunlight (usually morning) and plant chilli plants there.

REFERENCES

Agrawal, B. (2000) 'Conceptualising environmental collective action: Why gender matters', *Cambridge Journal of Economics*, vol 24, no 3, pp283–310
Anosike, V. (2004) 'Gender mainstreaming and women role in small scale vegetable production in a dynamic urban environment', paper presented at the RUAF/Urban Harvest Woman Feeding Cities Workshop, *Gender Mainstreaming in Urban Food Production and Food Security*, 20–23 September 2004, Accra, Ghana
Braun, B. (2005) 'Environmental issues: Writing a more-than-human urban geography', *Progress in Human Geography*, vol 29, no 5, pp635–650
Das, R. J. (2004) 'Social capital and poverty of the wage-labour class: Problems with the social capital theory', *Transactions of the Institute of British Geographers*, vol 29, no 1, pp27–45
Demerritt, D. (2002) 'What is the "social construction of nature"? Typology and sympathetic critique', *Progress in Human Geography*, vol 26, no 6, pp767–790
Espinosa, I. (2004) 'Perfil de género de la economía Nicaragüense en el nuevo contexto de la apertura comercial', UNIFEM, Managua, Nicaragua
Frayne, B. (2004) 'Migration and urban survival strategies in Windhoek, Namibia', *Geoforum*, vol 35, pp489–505
FUNDECI (2004) 'Informe Final de Memorial Sandino', unpublished report. Managua, Nicaragua
Guerrero, R. (2005) 'Huertos para reducir la pobreza en Managua', *La Prensa*, 29 July
Heynen, N., Kaika, M. and Swyngedouw, E. (2005) 'Introduction', in Heynen, N., Kaika, M. and Swyngedouw, E. (eds) *In the Nature of Cities*, Routledge, New York, pp1–20
Hovorka, A. (2005) 'Gender, commercial urban agriculture and urban food supply in Great Gaborone, Botswana', in Mougeot, L. J. A. (ed) *Agropolis: The Social, Political and Environmental Dimensions of Urban Agriculture*, pp137–152
Hovorka, A. (2006) 'The no.1 ladies poultry farm: feminist political ecology of urban agriculture in Botswana', *Gender Place and Culture*, vol 13, no 3, pp207–225
Instituto Nacional de Estadísticas y Censos (INEC) (2006) *Censo 2005: Volumen II Vivienda Municipios (VIII Censo de Población y IV de Vivienda)*, Managua, Nicaragua
Kabeer, N. (1994) *Reversed Realities: Gender Hierarchies in Development Thought*, Verso, London, UK

Lok, R. (1998) 'El huerto casero tropical tradicional en América Central', in Lok, R. (ed) *Huertos Caseros Tradicionales de América Central: Características, beneficios e importancia, desde un enfoque multidisciplinario*, pp7–28. CATIE, Turrialba, Costa Rica

McGuire, S. (1991) *Streets with No Names: A Journey into Central and South America*, Atlantic Monthly Press, New York

Méndez, V. E., Lok, R. and Somarriba, E. (2001). 'Interdisciplinary analysis of homegardens in Nicaragua: microzonation, plant use and socioeconomic importance', *Agroforestry Systems*, vol 51, pp85–96

Moser, C. (1993) *Gender Planning and Development: Theory, Practice and Training*, Routledge, New York

Page, B. (2002) 'Urban agriculture in Cameroon: An anti-politics machine in the making', *Geoforum*, vol 33, pp41–54

Peake, L. and Trotz, A. (1999) *Gender, Ethnicity and Place: Women and Identities in Guyana*, Routledge, New York

Rocheleau, D. (2001) 'Complex communities and relational webs. Uncertainty, surprise and transformation in Machakos', in Mehta, L., Leach, M. and Scoones, I. (eds) 'Environmental governance in an uncertain world', *IDS Bulletin*, vol 32, no 4, pp78–87

Rodgers, D. (2004) ' "Disembedding" the city: Crime, insecurity, and spatial organization in Managua, Nicaragua', *Environment and Urbanization*, vol 16, no 2, pp113–123

Silvey, R. and Elmhirst, R. (2003) 'Engendering social capital: Women workers and rural–urban networks in Indonesia's crisis', *World Development*, vol 31, no 5, pp865–879

Swyngedouw, E. (2005) 'Metabolic urbanization: The making of cyborg cities', in Heynen, N., Kaika, M. and Swyngedouw, E. (eds) *In the Nature of Cities*, Routledge, New York, pp21–40

Wall, D. (1996) 'City profile: Managua', *Cities*, vol 13, no 1, pp45–52

Wolch, J., Pincetl, S. and Pulido, L. (2001) 'Urban nature and the nature of urbanism', in Dear M. (ed) *From Chicago to LA: Making Sense of Urban Theory*, Sage Publications, Thousand Oaks, CA, pp367–402

12

Urban Agriculture and Physical Planning: A Case Study of Zaria, Nigeria

Chuo Adamu Nsangu[1]

INTRODUCTION

Zaria, located in Kaduna State, is a medium-sized city with an estimated population of 547,000 and a growth rate of 3.5 per cent per annum (Ministry of Economic Development, 1996). Its development and growth is attributed to its location in northern Nigeria and because historically Zaria has been a centre of learning with specialized training and research facilities. Agriculture is by far the most important activity of the working population. Approximately 40–75 per cent of Zaria's working population derive their principal means of livelihood from agriculture (ABU, 2000). Pressure on land tends to decrease as distance from the city centre increases – just as the percentage of people farming increases. The agricultural activity in Zaria can be divided into two types: rain-fed and irrigated farming.

The city is also part of an important watershed that divides the Sokoto and Chad River basins. The numerous rivers and streams found in this area include the Galma, Kubanni, Shika and Saye Rivers. The Galma River forms the main focus of the drainage system and carries water throughout the year; however, most of its tributaries dry up between January and June.

Zaria has a tropical climate with a mean total annual rainfall of approximately 1100 mm, and lies in the natural vegetation zone known as Northern Guinea Savannah that is primarily woodland. Soils in Zaria mostly belong to the class of leached, ferruginous tropical soils, whose material consists of several feet of deposited silt and sand overlying sedimentary decomposed rock. The soils of this group are heavy to work because they tend to be waterlogged with heavy rains; also they tend to dry out and crack during the dry season. In the 1930s, many regional institutions developed forest reserves and plantations for fuel supply.

Goodall (1978) states that the land-use pattern of an urban area represents the cumulative effect of myriad decisions and actions by various individuals and organizations. In Zaria, much of the urban space is used for agricultural practices. Urban agriculture has been of concern to physical planners, especially in developing countries, where the use of land for farming is common. Studies of the Nigerian cities of Lagos (Ezedinma and Chukuezi, 1999), of Ibadan (Tricaud, 1987) and Kano (Olofin and Tanko, 2003) have identified planning laws as a major constraint to UA. Legal constraints stem from various sources, including Nigeria's National Agenda 21, the National Policy on the Environment, the Land Use Act of 1978 (FRN, 1978) and Nigerian Urban and Regional Planning Decree (No 88) of 1992 (FRN, 1992).

The power to control development, as defined by these acts, has not been substantially debated or revised vis à vis UA largely because UA is a relatively new topic. Where reference is made, UA is banned outright. For example, the cultivation of annual and perennial crops, as well as the raising of livestock in urban areas, is not permitted under Nigerian law (Section 43, Land Use Act of 1978; FRN, 1978) except to preserve existing trees or plant new trees by the imposition of necessary conditions (Section 72, Decree No 88, of 1992; FRN, 1992).

These laws have generally made land rights and tenure difficult to secure, especially for the poor. Thus, urban laws and regulations require reform in order to improve access to UA (Tricaud, 1987; Ezedinma and Chukuezi, 1999; Olofin and Tanko, 2003). These concerns have been brought forward more often by agricultural policy makers and geographers than by town planners. Little is known, however, about how to effectively implement such reforms and how they will affect physical planning and city development. Therefore, the spatial implications of UA need to be studied in Zaria. The constraints imposed by existing legislation also need to be examined as a basis for improving knowledge about the practice of UA and its potential for enhancing sustainable development. This is the context that forms the basis of this study.

BACKGROUND, OBJECTIVES AND HYPOTHESIS

Part 1B of the Nigerian Urban and Regional Planning Decree (No 88) of 1992 outlines the administrative responsibilities of federal, state and local governments. According to the decree, the federal government formulates national policies for urban and regional planning, and prepares and implements national, physical, regional and subject plans. The national government also coordinates state and local governments in the implementation of their physical development plans, as well as providing technical assistance to the states. On the other hand, states are responsible for the development of urban and regional plans within their boundaries as well as for producing the state's development plans. States also control development of the land within their jurisdiction and conduct research in urban and regional planning. In contrast, municipal (local) governments are responsible for preparing and implementing town plans and for controlling development within their area of jurisdiction.

Section 1 of the Land Use Act of 1978 (FRN, 1978) notes that all land in each state is ultimately controlled by the governor of that state, who holds lands in trust and administers it for the use and common benefit of all Nigerians. The state governor is empowered to grant statutory rights of land occupancy to any person above the age of 21, for all purposes, regardless of whether or not the land is situated in an urban area (Section 5(4) of the 1978 Land Use Act).

Under Section 6.1 of this Act, the local government is given the power to grant customary rights of occupancy for the use of non-urban land for agricultural, residential and other purposes for a specific period of time. The Act does not define urban land or other land; however, governors are required to publish in the state gazette which areas in the state are designated as urban or as 'other' land. Section 43.3.4 of the Act notes that any person who contravenes any of its provisions 'would be guilty of an offence and liable on conviction to imprisonment for one year or to a fine of N5000.00 (US$42) or a fine of N100.00 (less than US$1 per day) each day during which he makes a default in complying with the requirements made by the governor'.

The Nigerian Urban and Regional Planning Decree of 1992 complements this 1978 Act. UA is not recognized except in Section 72, which relates to the preservation or planting of trees. In fact, because no state in Nigeria has officially recognized UA (Olofin and Tanko, 2003), UA can be considered a contravening activity.

In addition to there being no recognition for UA, state governors have imposed discretionary measures that have harmed its status and viability. For instance, in Kaduna State, the governor prohibited the use of government plots for any agricultural activity, despite the fact that public institutions own large areas of undeveloped land. This edict empowered the heads of public institutions throughout the state to ban farming at the institutions under their control. Zaria's Ahmadu Bello University destroyed crops on its premises in 1997 and 1998. Moreover, it completely banned the cultivation of tall crops on non-university experimental land or plots (i.e. on land within staff residential areas and on undeveloped land around the campuses of the university).

In general, planning constraints in Nigeria and in particular in Kaduna State have impeded the growth of UA and discouraged urban farmers in Zaria. Farmers do not invest in land because they are insecure about being able to maintain access and being able to farm the land in the near future. Thus, urban farmers hardly ever obtain the maximum production from their farms.

This research examines the pattern and implications of agriculture in Zaria and provides recommendations for physical planning. The study had four objectives:

1   Assess urban land use policies relating to urban agricultural development
2   Determine the pattern and characteristics of UA in Zaria
3   Determine implications of UA on physical planning in Zaria
4   Make recommendations for physical planning.

The following three research questions were examined:

1   What patterns of UA have emerged in Zaria?
2   What implications does UA have for physical planning in Zaria?
3   What planning reforms are needed to develop UA in Zaria?

The following hypothesis was proposed:

> The amount of land devoted to UA in Zaria is related to the allocation of urban land use, where the ratio of land development is expressed as the area of actual development per plot size to the rate at which urban land has been developed.

METHODOLOGY

A total of 200 farmers from Zaria who represented a range of different income groups were chosen for the study. Farmers all had access to land and were directly involved in UA. Data were collected from questionnaires, interviews and satellite images. The latter were interpreted to determine land uses, which were then cross-referenced to using a survey-grade global positioning system (GPS). Questionnaires collected information from the sample, and interviews were conducted with politicians, heads of planning agencies, urban managers, farmers and the general public.

For the field survey, assistants distributed questionnaires to farmers throughout Zaria's urban districts. The survey was designed to collect information on household characteristics, farm scale, farm produce, access to land and farmers' desires or preferences in relation to farming. Fifty people were randomly selected from each district and every person provided their responses to the field assistants. Heads of planning agencies or units, urban managers, politicians, urban farmers and the general public were also interviewed in order to understand their opinions on and ideas for managing UA. Statistical tests including analysis of variance, the *f-test* and *t-test* were used to test the hypothesis and analyse findings related to the distribution and extent of UA vis à vis developed land in the town.

The research report was presented to top-ranking members of the Nigerian Institute of Town Planners, who evaluated the work in both internal and external examinations.

RESEARCH FINDINGS

Agricultural activities in Zaria

Horticulture was found to be the dominant agricultural activity on public and semi-public lands, as well as in low-density residential areas. Aquaculture was

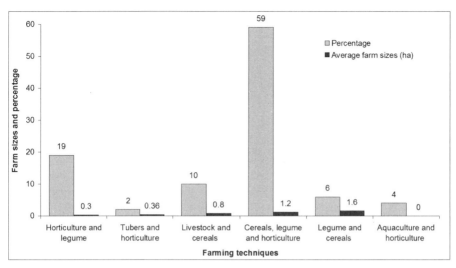

Figure 12.1 *Distribution of farm sizes and techniques in Zaria's urban area*

undertaken in water reservoirs and major rivers and livestock was raised largely on the outskirts of the city. However, cereal crops and legumes were grown everywhere except where agricultural controls were strictly enforced.

In Zaria, urban farmers typically cultivated more than one crop, as shown in Figure 12.1. The figure also shows the various size of the farm in relation to the crops grown. Approximately 60 per cent of the respondents planted cereals and legumes and also practised horticulture, whereas only about 2 per cent planted tubers and also practised horticulture.

Cereals are a staple food item in northern Nigeria as well as in the savannah, which accounts for their popularity with farmers. The frequency of horticulture among urban farmers can be attributed to natural factors such as climate, soils and the availability of floodplains. In addition, many low- and medium-density residential areas in Zaria have large open spaces, which are suitable for horticulture. Consumer demand was found to be an important factor in the farming practices in Nigeria.

Desire for urban space for agriculture

Respondents were asked whether they would like to have more space for farming and what would be their preferred way of farming. In Zaria, urban farmers expressed a strong desire to farm and stated that they will probably continue to do so in the future. Many farmers see UA as a critical part of their livelihoods and as a long-term livelihood strategy. Their interest in it is attributed to inconsistent food prices or to the poor system of food distribution in Nigeria. Therefore, farming helps ensure food security in those areas where income levels are low and while inflation is increasing. Other reasons for its popularity include cheap and available labour, access to agricultural inputs, as well as easy access to markets where produce is sold –

Table 12.1 *Desired farming arrangements in Zaria*

	No of samples	Location preference (%)		Agricultural technique preference (%)			
		Urban core	Peri-urban	Horticulture	Legumes	Cereals	Livestock
0.1–1.0	32 (16%)	100	–	15	25	60	–
1.1–2.0	70 (35%)	67	33	17	20	53	10
2.1–3.0	48 (24%)	25	75	17	20	41	22
3.1–4.0	26 (13%)	–	100	–	15	65	20
4.1–5.0	24 (12%)	–	100	–	35	60	5
Total	200 (100%)	38.4	61.6	9.8	23	55.8	11.4

Source: Field work, 2001.

which leads to fewer transportation costs. Farmers' preferences for engaging in agriculture within urban or peri-urban areas, in addition to their expressed desire for a variety of farm sizes (Table 12.1), suggest how UA could affect the urban landscape of Zaria.

Most farmers in Zaria are commuters who live in the city and commute to their farms in the peri-urban areas during farming seasons. They do not have an actual farmstead or herds to manage which would necessitate them migrating to and living in the area during the farming season. Some farmers acquired land by renting it, usually for one farming season at a time. And although renting land was seen as theoretically feasible, in terms of land security it is weak because leases are not always renewed by landlords, who will not rent to the same person again if they receive a better offer.

A typology of the spatial patterns of urban agriculture

As with most cities, UA in Zaria is commonly employed on different types of vacant land regardless of whether it is private, public or semi-public. UA is also practised in other areas where access to land is easy such as in commercial areas, industrial areas, along rights-of-way, floodplains and other open spaces. This project prepared a typology of UA land uses which includes the following:

• *Urban agriculture on public and semi-public land:* Agricultural activities in any part of the township depend on how much of the land has been developed and upon the control measures adopted by those institutions owning the land. Most public and semi-public lands are farmed. The institutions that own the land include the Nigerian Railway Corporation, Ahmadu Bello University, the Nigeria Police Academy, the army depot, the Nigerian College of Aviation Technology, the Federal College of Education, military barracks, missionary areas, state polytechnics and secondary schools and other institutions. The dominant crops are cereals, legumes and tubers, but horticulture is also practised.

- *Urban agriculture in residential areas:* Cultivation takes place in many low-density residential areas in Zaria. The plot sizes in most low-density residential areas are 65 m × 65 m. The residential areas of low- to medium-density are also cultivated (such as Hanwa and the northern part of Tudun Wada extending along the Wusasa, Kwangila and Paladan axis). In addition to cereals and horticulture here, there are pockets where livestock (especially goats, sheep, birds and, in some districts, pigs) are raised within the residential areas.
- *Urban agriculture in commercial areas:* The proposed city centre of Zaria is yet to be developed. Over the years, farming has occurred on some of the land prescribed for urban development. In addition, residential developments have been built on 14 per cent of the land. Other areas of commercial use proposed in the master plan have not been developed and, in some cases, urban farmers have taken advantage of this vacant space.
- *Urban agriculture in industrial areas:* Several industrial zones in the city have vacant lands that are being cultivated, including the light industrial zone of Chikaji in the Sabon Gari District and the heavy industrial zone along Jos Road (Figure 12.2). The emergence of agriculture in these areas can be attributed to the large availability of vacant land. In addition, the economic recession in Nigeria has resulted in less investment in the industrial sector. Land originally for industrial purposes is often either not developed or only partially developed, increasing the opportunities for UA.
- *Urban agriculture along rights-of-way and other open spaces:* Zaria has a large network of railways and roads. The Sabon Gari district has the most rail and road rights-of-way and these – especially those that pass through Tudun Wada, the government reservation area (GRA), Sabon Gari, Kano and Gusau – are used for agricultural activities. The rights-of-way under utility lines such as electrical lines, as well as the access routes to the junctions, are also used for farming. Even the Kano Road junction at Kwangila, where the Sokoto national grid and the Kano national grid are linked, is used for farming. This is probably because the utility lines pass through floodplains or swampy areas and farmland is otherwise not accessible.
- *Urban agriculture on floodplains:* Soil quality affects the distribution of agricultural production. Production varies according to differences in the physical, chemical and biological characteristics of the soil. In Zaria, the most productive agricultural land is found along the floodplains of the Kubanni and Galma Rivers. The soils are hydromorphic, and as irrigation water is available, farming is a year-round practice. Crop distribution patterns are haphazard because of the types of land tenure, which include freehold, tenancy and ownership. Land acquisition patterns tend to lead to small individual plots. Cereals are the dominant crop produced on the floodplains. The second most prevalent agricultural activity is horticulture, particularly in irrigated areas. In this category, mostly vegetables are produced, but in some cases, fruits are sandwiched among cereal crops in upland areas. Tubers, especially cassava and cocoyams (taro), are cultivated upstream of the floodplains.

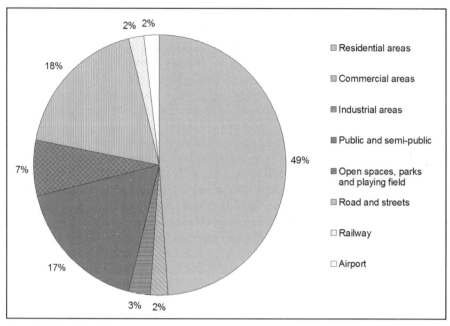

Figure 12.2 *Urban land use in Zaria (percentage)*

- *Peri-urban agriculture (UPA):* The periphery of the city is more intensively and extensively farmed than core urban areas. Here, farms are larger and more economically sustainable than the small farms located within the urban areas. However, UPA is not controlled or regulated by the authorities. The choice both of crop and production process is determined by the farmer and market forces. Cereals, legumes and livestock, however, still tend to dominate the farming systems in this area.

To arrive at estimates of the extent of UA, the land-use map of Zaria was used along with GIS techniques during the research. The researcher merged the proposed land-use map with a satellite image of Zaria Township and digitized those areas devoted to UA and other uses. Figure 12.2 shows existing land uses in the township of Zaria. An estimated 58 per cent of the total land proposed for 'other uses' is presently being used for agriculture (Table 12.2). Physical planning problems – constraints – and the slow rate of economic growth may have increased the amount of land devoted to agriculture instead of the planned use.

With respect to how farmers have acquired land, Table 12.3 shows that 28 per cent of urban farmers in Zaria inherited their farmland or received it as a gift, 44 per cent rented from a landlord, and 21 per cent leased the land from public or semi-public institutions. An additional 7 per cent of respondents had converted a building plot to a farm. The process of informal land acquisition does not provide a platform or any formal legal security for urban farmers.

Table 12.2 *Urban land devoted for agriculture in Zaria*

Urban land use	Land devoted for agric (ha)	Percentage of land currently used for UA
Residential areas	7316	54
Commercial areas	399	50
Industrial areas	534	67
Public and semi-public spaces	2668	58
Open spaces, parks and playfield	1061	56
Roads and streets	2809	65
Railways	268	63
Airports	300	47
Total	15,357.7	58 (mean)

Source: Urban field work, 2001.

This disadvantage has prevented them from forming cooperative bodies, political and social groups that are formally recognized by the authorities. Also, it has hindered farmers from negotiating for technological improvements and asking for financial and credit assistance from the government or private-sector banks. In fact, UA is not recognized by the government. The ramifications of this create a great deal of uncertainty about the future of UA in Zaria.

Determinants of urban agricultural patterns in Zaria

As Table 12.2 shows, much of the land proposed in the master plan for Zaria's urban area is being used for agriculture. The development of agriculture has been encouraged by the availability of land (under many different jurisdictions) in the urban area, the abundance of incidental vacant land, socio-economic factors and so on.

Availability of land

Most of the land in Zaria's urban area has remained undeveloped regardless of its proposed use. As a result, these lands become attractive areas for

Table 12.3 *Mode of land acquisition for urban agriculture in Zaria*

Mode	No of respondents	Percentages
Inheritance/gift	56	28
Rent/hire	89	44
Leasehold	41	21
Conversion of a building plot to a farm	14	7
Total	200	100

Source: Urban field work, 2001.

Table 12.4 *Pattern of land development in Zaria*

Urban land use	Proposed land (ha)	Existing development in year 2000	Percentage developed as proposed (as of 2000)
Residential	13,302	5320	40
Commercial	798	319	40
Industrial	798	183	23
Public and semi-public	4522	1492	33
Parks, playfield/open spaces	1862	595	32
Roads and streets	4256	1149	27
Railway	425	119	28
Airport	638	236	37
Total	26,604	9417	32

Source: Zaria Master Plan and Field Work, 2001.

agriculture. Table 12.4 indicates that only 32 per cent of the total land projected in 1975 to be developed by the year 2000 was actually developed. Town planners may not have been able to accurately estimate Zaria's growth because they had a poor understanding of economic growth in the town – as well as of the town's role in the national economy. Nor did political decisions allocating land for particular activities consider what would happen in the event of non-development.

In actual fact, development control measures and zoning regulations seem to have inadvertently paved the way for UA in Zaria. For instance, the government has not released funds for development of townships as was proposed in the Zaria master plan. Therefore, much land zoned for specific uses has yet to be converted to those proposed. Development control measures became ineffective because there remains an anticipation that one day the government may release funds for the proposals. In other words, non-implementation of the government's master plan means agriculture is left to proliferate.

An analysis of variance showed the proposed land allocation for various urban uses contributed to 97 per cent of the agricultural development in the town. This figure regarding farmlands was affected by the amount of open space within farmland that was intended for road access and utility lines as prescribed by the master plan. This is in contrast to the amount of actual developed land as specified by the master plan for Zaria's urban area.

Using the f-test, the calculated f statistic of 77.93 indicated that at least one of these variables had a significant contribution to the development of UA in Zaria. This is because development has only been carried out on a small fraction of the land allocated for development. The calculated value was much larger than the expected f statistic: $f(3,4) = 16.69(P \leq 0.01)$. The t-test was used to establish the contribution of each variable. The proposed land

Table 12.5 *Proposed urban land uses vs actual land developed in Zaria*

Urban land use	Land devoted to agriculture = Y	Proposed land = X_1	Open spaces = X_2	Actual land developed = X_3
Residential	7316	13,302	5320	5320
Commercial	399	798	143	319
Industrial	534	798	103	183
Public and semi-public	2668	4522	1356	1492
Parks and open spaces	1061	1862	93	595
Roads and streets	2809	4256	2554	1149
Railway	268	425	55	119
Airport	300	4522	1356	236

Source: Urban field work, 2001.

allocation and open space were shown to have contributed somewhat; however, the contribution from the actual land developed was the most significant, with $t(4) = 2.77(P \leq 0.05)$.

The above analysis implies that the physical-planning proposals did not meet the planning goals and objectives of the town, which may be attributed to poor predictions of development within Zaria. If development does not occur, UA fills the void.

Abundance of incidental vacant land

As an activity that is commonly associated with the poor, UA is often conducted on marginal lands. In Zaria, UA is commonly found in the floodplains, which are exposed to natural hazards. Two major rivers, the Kubanni and Galma, largely drain Zaria's urban area and UA is found on the floodplains of a number of tributaries of these rivers. Productivity is encouraged by the hydromorphic soils. In addition, water is available for most of the year, which enables land to be irrigated in the dry season. Apart from the floodplains, some isolated pockets of land exist within the town that cannot be used for development because of their size or location. Moreover, little is done to regulate activities on the floodplains, which leaves farmers to practise their livelihoods.

Influence of socio-economic factors

Employment type and the standard of living influence what kinds of part-time activities are required to earn a living. The high participation of people in UA is partly the result of a difficult economic situation. It is not just a hobby. Both employed and unemployed people actively engage in urban farming, as shown in Table 12.6.

As Table 12.6 shows, 34 per cent of the respondents involved in UA were unemployed and an additional 12 per cent of respondents were retired. However, 33 per cent of respondents were working in either the public or private sector, while 21 per cent were self-employed.

Table 12.6 *Occupational status of urban farmers surveyed in Zaria*

Occupation	Number of respondents	Percentage
Public and private sector	66	33
Self-employed	42	21
Retired workers	24	12
Unemployed	68	34
Total	200	100

Source: Urban field work, 2001.

Table 12.7 shows how much farm produce is consumed and sold by urban farmers. Farmers and their families consume approximately 81 per cent of their own produce and sell the rest. This situation – where UA is used primarily for sustenance – may be attributed to a poor food distribution system, as well as an increase in food prices. In addition, when families are large, there is more food consumption, so only a little produce can be sold to acquire farm inputs or cooking ingredients.

Implications of planning laws and regulations on urban agriculture

Physical-planning laws in Nigeria ignore UA. Thus, despite the potential benefits, UA is not recognized as an important activity. In fact, development-control measures often have a negative impact on the development of agriculture. Some of the problems include reductions in farm size (and thus a reduction in urban farmers' income) and an increased food insecurity in Zaria.

Table 12.7 *Distribution of farm produce consumed and sold by urban farmers in Zaria*

Age group (years)	Number of respondents	Average family size	Annual percentage of own produce consumed	Annual percentage of own produce sold
28–33	24	3	75	25
34–39	32	5	77	23
40–45	54	7	70	30
46–51	38	9	86	14
52–57	20	12	81	19
58–63	18	13	79	21
64–69	14	12	88	12
Total	200	7.6	80.7	19.3

Source: Field work, 2001.

Table 12.8 *Effect on farm sizes in Zaria of planning laws and regulations*

Farm area lost (ha)	Rate of income lost (in %)	Rate of drop in annual output (% of total land lost)	No of respondents	Percentage
0.10–0.2	5–10	10–15	22	11.0
0.21–0.4	11–15	16–20	27	13.5
0.41–0.6	16–20	21–25	35	17.5
0.61–0.8	21–25	26–30	41	20.5
0.81–1.0	26–30	31–40	56	28.0
No effect	No effect	No effect	19	9.5
Total			200	100

Source: Urban field work, 2001.

Table 12.8 shows that 28 per cent of the farmers report a reduction in the size of their farm in recent years related to planning controls and municipal by-laws. They also report a corresponding loss of income and drop in annual output. However, 11 per cent of the farmers registered few losses. Only 9.5 per cent of the farmers had not registered any loss of their farm sizes, income and output due to planning laws and regulations. The table also depicts a direct, inverse relationship between reduction of farm size, income loss and reduction of annual output, respectively. In other words, the reported small loss in farm sizes is associated with both a small loss of income and small drop in output. Alternatively, the large drop in output is associated with high income-loss and large reduction of farm size.

Planning controls have influenced farm sizes through the encouragement of development in areas that were once used for farming. Where physical development has been slow and there is less enforcement of controls over development, farm sizes are not affected and agriculture prevails. This situation becomes reversed where development is fast and where controls upon it are strictly enforced. In such cases, sizes of farms tend to be drastically reduced.

The annual income lost because of physical-planning control measures was also estimated, as indicated in Table 12.8. To date, farmers had made claims of lost income because they were not allowed to make any decisions on the type of crops to cultivate. In some cases, the type of crops they were allowed to grow failed to produce. Results could include a drop in their income or they might discover that the crops they were permitted to grow were not the type which could reliably generate an income.

Control measures and bans on farming in certain areas have also significantly affected the annual output of urban farmers, as indicated in Table 12.8. For example, the use of guinea corn and maize stems for fencing is prohibited in some places, which encourages pedestrian traffic through the crops. Also, restrictions on the types of crops to cultivate compromises the production of potential crops that might perform better if only farmers were

Table 12.9 *Mean and potential yield estimates of cereals on uncultivated (underutilized) land*

Crop	Mean yield (t/ha)*	Potential yield (t/ha)*	Mean yield estimate for 250 ha (t)**	Potential yield estimate for 250 ha (t)**
Upland rice	0.8–1.2	1.5–2.5	250	500
Lowland rice	1.0–2.0	2.5–8.0	375	1312.5
Maize	1.5–2.1	3.5–10.0	437.5	1687.5
Sorghum	0.5–1.2	2.0–2.5	212.5	562.5
Total	–	–	1275	4062.5

Source: *NCRI Survey, 1988.
**Assumption made based on field work, 2001.

permitted to make their own decisions on the type of crop to cultivate and where to do so. This is because crop yields vary greatly, being tied to fertility levels of soil. A decrease in production is also associated with the theft of crops, as well as with their destruction by roaming animals.

Finally, planning controls have significantly affected cereal production in Zaria's urban area. Cultivation has been completely banned in some places, whereas only tall crops have been banned in others. In addition, farming has been banned on approximately 250 ha of arable land in Zaria's urban area. Table 12.9 illustrates that when the average yield of cereal crops on 250 ha in Zaria is compared with the potential yield, the ban on a 250-ha area creates an annual loss of 1275.0–4062.5 tonnes of cereals. This production loss naturally affects the total food supply of the city.

Constraints to urban agriculture – and farmers' coping strategies

Despite the effects of physical-planning controls and regulations on UA in Zaria, urban farmers have adopted various coping methods. As shown in Figure 12.3, 44 per cent of respondents indicated they were involved in part-time businesses, 22 per cent restricted their sale of produce, whereas 5 per cent relied on assistance from friends and relatives. However, 6 per cent of the respondents indicated that they did not have any other options to supplement their family diets.

Table 12.10 shows an overview of the attitudes of different stakeholders towards UA. Each was assessed by the researchers as to their general level of sympathy and support for UA. Government agencies have proven to be indecisive in providing land or protecting users' rights on land used for agriculture. However, when the negative implications of UA on society become severe, physical planners attempt to control UA. On the other hand, society in general remains largely sympathetic to UA. People support farmers

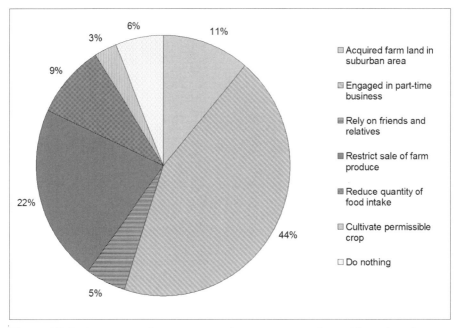

Figure 12.3 *Coping mechanisms regarding strategies adopted by urban farmers in Zaria as influenced by physical-planning control*

and also they often harbour doubts about government intentions and planning. Therefore, this research shows that UA advocates require the support of scientific proof in order to debunk the claims of government bureaucrats who believe UA has negative implications for urban life in this part of the world.

RECOMMENDATIONS AND CONCLUSIONS

Urban agriculture has survived in Zaria because of various interactions within urban areas. Factors supporting farming practices include: natural forces that make urban land (especially floodplains) available for agricultural development; limitations of the planning process which encourage agricultural opportunities; and compatibilities between UA and certain areas of industrial use, such as open spaces and, in some cases, public and semi-public land within urban areas.

Such circumstances require a review of the legal framework so that agricultural activities that have minimal negative impact can be better accommodated. In particular, Section 72 of Decree No 88 of 1992 needs to be amended with respect to the Harare Declaration. In 2003, the Harare Declaration was signed by the Ministers responsible for local government from Kenya, Malawi, Swaziland, Tanzania and Zimbabwe. The Declaration

Table 12.10 *Evaluation of stakeholder concerns regarding UA in Zaria*

Constituent	Employment opportunity and income generation	Access to land and users' right	Food security	Safety and security	Adverse effect on health	Adverse effect on environment	Physical planning constraints	Adverse effect on town aesthetic	Land speculation	Formation of urban agriculture cooperatives	Means of waste management
Sabon Gari LGA	PS	I	PS	N	S	S	I	S	S	I	PS
Zaria LGA	PS	I	PS	N	S	S	I	S	S	I	PS
Federal ministry of works and housing	S	I	S	N	I	I	I	N	N	N	I
Planning units within government institution	S	I	S	S	N	I	S	N	I	I	PS
Environmental agencies	S	I	PS	S	S	I	S	N	I	I	S
Health agencies	S	I	I	PS	N	I	I	I	I	I	I
Physical planners	S	I	PS	N	N	I	I	N	N	I	PS
Farmers	PS	PS	S	PS	I	I	N	S	I	PS	PS
General public	S	I	PS	S	S	S	N	I	I	I	PS
Politicians	S	S	PS	S	S	I	S	S	S	I	PS

PS, positive and supportive; S, sympathizers; N, negative; I, indecisive.
Source: Urban field interview, 2001.

officially endorses UA as a development strategy. This declaration should be seen as an inspiration for the development of UA in Zaria and Nigeria as a whole.

The research presented in this paper showed that UA was found to have adverse implications in some areas of Zaria – these require planning intervention. Ideally, the most appropriate method for identifying potential agricultural areas requires an agricultural land-use map. The map would classify agricultural areas, ranging from prime agriculture land to the least productive land. Continuous pieces of land in a town of high agricultural productivity may not be geographically possible; therefore, urban practitioners need to be flexible and exploit areas that are potentially favourable for agricultural development.

Proposed location requirements for urban agriculture

Urban agriculture consists of the cultivation of crops that vary in the amount of space they require. Therefore, locations for UA should be large enough to accommodate the growth and development of crops. In the case of livestock, space must be sufficient for the animals to forage or to be raised in a fenced, gated pastures. Also, agricultural activities entail a host of inputs and outputs. Transportation, labour – and proximity to labour – tools and other means of production, market access, land and water are all important variables related to UA. Proximity to and access to these means of production can be aided (or impeded) by land-use plans. Urban farming also benefits from access to infrastructure: water provided by pipe; sewage for use as crop and forage manure; sewerage for waste disposal; and power for heating, cooling, lighting and production processes.

Based on this research, some guidelines are proposed for any future policy thinking on UA in Nigeria:

- Farming should be prohibited on lands prone to erosion.
- Farming should be prohibited in densely populated areas.
- Farming should be prohibited on rights-of-way and under utility lines.
- Farming activities that result in health risks and environmental degradation should be prohibited.
- Farmers should be monitored during application of chemicals and only specified ones should be applied to crops.
- Farming along floodplains should have adequate setbacks from the river channel or banks.

Further studies are required in the following areas: UA and urban growth, as well as the economic competition for urban space and changes in development policy. In addition, a cost-benefit analysis of UA should be conducted.

Major setbacks to this study included the lack of a UA land-use map with classifications and a lack of formal recognition of the usefulness of UA. Both are required for the proper design and management of Zaria's urban area.

ACKNOWLEDGEMENTS

I want to thank Canada's International Development Research Centre (IDRC) for its generous support through its AGROPOLIS award programme. IDRC financed this project and provided relevant literature for my studies. I greatly appreciate the Federal Department of Agricultural Land Resources GIS Unit – Livestock House, Mando Kaduna State, Nigeria – for providing GIS software and technical staff to assist in analysis. Also, I am grateful to my supervisors, Dr Ahmed Adamu and Mallam Usman Abubakar Siddiq of the Department of Urban and Regional Planning, Ahmadu Bello University, Zaria, who guided me throughout the research.

NOTE

1 Chuo Adamu Nsangu, MSc, Urban and Regional Planning, Mambim, Babessi Subdivision, Ngoketunjia Division, Northwest Province, Cameroon, tel: +237 75127389, email: achuo1@yahoo.com

REFERENCES

Ahmadu Bello University (ABU) (2000) *Zaria Master Plan*, Department of Urban and Regional Planning, ABU, Zaria, Nigeria

Ezedinma, C. and Chukuezi, C. (1999) 'A comparative analysis of urban agricultural enterprises in Lagos and Port Harcourt, Nigeria', *Journal of Environment and Urbanization*, vol 11, no 2, pp135–144, available at http://eau.sagepub.com/cgi/reprint/11/2/135.pdf (accessed 29 March 2008)

Federal Republic of Nigeria (FRN) (1978) *Land Use Act*, FRN, Lagos, Nigeria

Federal Republic of Nigeria (FRN) (1992) *Nigerian Urban and Regional Planning Decree* (No 88), Supplement of *Official Gazette Extraordinary* No 75, 31 December 1992. FRN, Lagos, Nigeria

Goodall, B. (1978) *The Economics of Urban Areas*, Pergamon Press, New York, NY, USA

Ministry of Economic Development (1996) *Kaduna State Statistical Yearbook, Kaduna, Nigeria*, Ministry of Economic Development, Statistics Division

National Cereals Research Institute (NCRI) (1988) Survey, NCRI, Badeggi, Nigeria

Olofin, E. A. and Tanko, A. I. (2003) 'Optimising agricultural land use in Kano', *Urban Agricultural Magazine*, No 11, December 2003

Tricaud, P. M. (1987) 'Urban agriculture in Ibadan and Freetown', *Food-Energy Nexus Programme Report 23*, United Nations University, Tokyo, Japan

13

Conclusion

Mark Redwood

In the face of urban development and the stunning rise of value in urban land markets, urban agriculture (UA) continues on. Although it is not an activity that will ever make huge financial sums, it is a keystone in the lives of millions of people in the South. The argument that UA will inevitably be replaced by buildings and other built infrastructure is a fallacy: UA has always existed and will continue to do so. Moreover, as we reconsider how the sustainable city of the 21st century will evolve, the inherent promise of UA to reuse wastes and minimize the environmental impact of food production will be a significant part of this visioning process.

UA will never replace or compensate for rural agriculture, but should be seen as a livelihood that enhances food security, nutritional health, and creates employment. It can also be both a permanent feature of a city or a temporary relief for a crisis situation. UA does not necessarily mean community or public management; it is an activity that can be managed privately, on private land. In fact, in the face of ever increasing populations and changes in food markets, UA will see an increase in its relative importance. At the end of the day, however, UA will not survive without protection from planners from development interests. This means cities can offer temporary occupancy permits, appropriate zoning of specific lands for agriculture, support for food markets, encouragement of farmers' organizations, and extension or training for farmers. Experience shows that a simple package of policies can result in major change.

Researchers and scientists have done their job: UA is now firmly on the development agenda. This book is only a small drop in a large body of work currently available. It is now up to international agencies, donors and, most importantly, national and local governments to take up the challenge of implementing programmes and support for UA.

This book presents UA as a crucial part of the equation that will lead cities to a state of food security. Urban food security is about increasing people's access to food, increasing its affordability and improving marketing and health-related practices to ensure that food is safe for everyone to eat. UA contributes by eliminating costly and inefficient transportation from rural areas, providing vegetables and nutritious foods grown locally where they are consumed and preserving some employment opportunities for those involved in raising such food.

The case studies illustrated how urban farmers are providing a service not only to the urban economy in terms of food production, but also in terms of mitigating some of the negative impacts of waste. For instance, UA manages to reuse wastes, particularly those which are organic. It is notable that even where waste reuse is common, an estimated 80 per cent of nutrients escape into the waste stream without being used. The provision of such ecosystem 'services' has ramifications for the population and also reduces costs associated with municipal waste management.

Proliferation of data and research on urban agriculture

The AGROPOLIS programme endeavoured to support the generation of data and research activity on UA. Now, ten years on, we are seeing a number of universities develop courses and research programmes on UA and integrate them into their curriculum. Ryerson University in Toronto has launched a distance-learning course on UA as part of its work on food security; the School of Architecture at McGill University conducted studios on UA under the guide of the successful 'Making the Edible Landscape' project; the University of Georgia (USA) now hosts the Georgia Center for Urban Agriculture in the Faculty of Agriculture; and Xavier University College of Agriculture in the Philippines is only one of many universities now teaching and conducting research on UA.

What is perhaps the most striking note regarding the rise of UA is how quickly interest has grown in development literature over the past 25 years. Figure 13.1 illustrates this rapid rise. The data show the number of peer-review articles *specifically* on UA over four six-year periods since 1985. Between 1985 and 1990, six articles were found in several major databases of abstracts. That number rises to a remarkable 110 peer-review articles between 2003 and 2008. This does not take into account articles that may be on related subjects nor the grey literature on UA.

In fact, more information than ever before is now available regarding this topic. As illustrated in the introduction, advances in institutions such as the Food and Agriculture Organization (FAO), UN-HABITAT and in universities have helped the proliferation of work in the area. This is because it is a topic that resonates. High-profile action such as the celebrated case of Rosario in Argentina, Kampala in Africa and the inclusion of UA in Brazil's FomeZero (Zero Hunger) programme have raised its profile. But the biggest success is that urban farmers, once marginalized in their cities, are finding their voice and forming groups to protect their rights. These farmers are not alone in their fight for recognition. However, notwithstanding such impressive gains, many more questions remain.

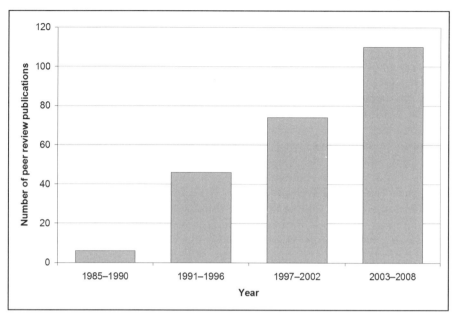

Source: CAB Abstracts and academic search premier.

Figure 13.1 *The growth of peer-reviewed articles with 'UA' in their title, 1985–2008*

Some remaining questions for researchers

Economic issues

At the centre of UA is a question that has not yet been satisfactorily answered: precisely what is the contribution that UA makes to the urban economy? Many books, articles and research begin with the premise that UA is making a contribution. However, a question remains: how much of an economic influence does it actually have? How are profits spent? Who benefits directly or indirectly?

Much of the work exploring UA through the lens of economics touches only on descriptive financial data. Data on revenue and costs are acquired and presented over relatively short periods. To date, there have been few integrated, large-scale economic studies on the topic. Many are limited surveys of a few informal farmers, so there is not much in the way of comparative methodologies. Nonetheless, most reports on UA start from the premise that it is an important economic activity which only has case evidence to offer as proof. In order to confirm this statement, it is important that robust, broad economic analysis be conducted.

Another issue that confounds researchers is the scale at which UA takes place. UA exists on a small scale (e.g. household subsistence gardening) as well as on much larger scales (livestock farms and peri-urban agricultural industries). Revenue generation in the former may simply be in the tens of dollars per month, whereas the financial gains in the latter might be in the tens of thousands of dollars per month. Smaller-scale farming can be of profound importance to an individual household's economy, while the latter is making a significant contribution to the community as a whole.

The question strengthens the claim that producers urgently need better access to credit if they are to increase their production efficiency such that women – who are often very active either as farmers or marketers – can improve their access to financial assets. Despite this perhaps obvious conclusion, there have been few direct and strong economic analyses conducted related to the value of UA. Moreover, such analyses must be corroborated with the economic value generated through other uses of the land (such as housing development).

The same can be said of urban land markets. Mara and Feachem (2001) have famously surmised that the poor occupy land valued at US$9 trillion worldwide. This value is locked up since the very same poor often occupy their lands illegally or with limited tenure. What would happen if there was a semblance of security created so that this land could be used to leverage grants, credit and access to the formal banking systems and agricultural support? The issue of credit as it relates to investment in UA could feasibly dominate discussions about how this sector will be developed in the future.

Pollution management

The effects of pollution and the relationship between UA and the environment are much better understood than they were ten years ago. The work of many international organizations – notably, the Consultative Group on International Agricultural Research (CGIAR) system, including International Development Research Centre (IDRC) partners – has led to a large body of work on waste management and UA. Readers will note this book includes several chapters on wastewater and one on solid waste. These are important and well-explored topics related to UA and are studied for good reason, primarily because management options for water and solid waste are so closely related to urban farming. These wastes hold enormous monetary value (Dreschel et al, 2007) and reflection upon them challenges us to ask: How can we optimize the capture of these valuable resources?

Other types of pollution are only occasionally considered. Cars are an inescapable fact of the 21st century city – however, with few exceptions, air pollution is considered not more than an irritant and few people explore the direct links between air quality and automobiles. In fact, evidence exists that air pollution has a negative effect on crop yields (Agrawal et al, 2003). However, such definite links are far from adequate enough to convince policy makers to control vehicles or manage emissions. It is important that research be initiated to explore the link between vehicular emissions and the food chain, which is associated with the growing work on air pollution and its impact on human health. Likewise, although a body of work on pesticides, organic chemical pollutants and urban food-production exists, the response to this potentially significant threat is negligible. The point here is that research can break new ground, but often a body of work needs to be large, multi-regional and accepted by many before it has the power to change policy and alter practice.

Health and the 'farm to fork' approach
Health-related problems are an important impediment to the uptake of UA but are not impossible to manage. In their programme on agriculture and cities, the International Water Management Institute (IWMI, 2007) advocates a multi-barrier approach to risk. For instance, tracking produce from 'farm to fork' and ensuring that at each stage there are simple barriers in place along the production–consumption chain (e.g. hygiene, gloves, cleaning of foods and markets, advertising) to alert and educate people (Amoah et al, 2007). Farmers would have one set of strategies to reduce risk, market vendors another, consumers still another and so on. Municipal action at the market level is perhaps the most effective: markets are geographically discrete, they are usually managed by the city, and it is the space where consumers and sellers come together. This means outreach can have a great deal of impact.

Understanding the peri-urban reality
Recent work on peri-urban areas reveals that resource flows create a reciprocal back and forth between cities and their surroundings. As work on UA grows, so too do the different interpretations of what 'urban' actually *means* vis-à-vis what happens in the area surrounding the city, that is, the peri-urban area. Recent work on common property, environment and development theory also has raised the importance of the peri-urban interface in our understanding of the city. The notion of a city's 'ecological footprint' suggests that the environmental – and, of course, the economic – impact of a city extends far beyond the reach of its administrative boundaries. Geographically, the peri-urban region can extend as much as 100 km from the centre of a city. It is here where food production for the urban market is concentrated, and this area exhibits characteristics common to both rural and urban areas.

With an increasing amount of recognition of the 'peri-urban' as having significant impact on the city, it is impossible for city planners to ignore it – hence the rise of regional authorities and regional planning. Since these areas are often outside of municipal control (because cities outgrow their original administrative boundary), the urgency to engage national and provincial authorities is acute. Indeed, the pressure of urban development on peri-urban greenfield sites is a serious development challenge.

Policy change
Policy makers need to first consider their primary policy objective. Is it to reduce poverty or to increase food availability? Is it to raise incomes of the poor? Or is it to increase green space and options for reuse of wastes? Such questions represent only the tip of the iceberg.

There are general lessons which we have come to understand regarding policy development for UA. The central lesson is that policy must be incentive based and not punitive. Incentive structures which help to encourage good practices and discourage bad ones are critical but, unfortunately, rare. Most often, 'bad' practices are punished without any other feasible options being put forward. Central to this is the acknowledgement, once and for all, that UA is an active part of the city economy. The case studies provide some insight into existing policies as cities wrestle with the pros and cons of UA. The case of Zaria, Nigeria (Nsangu) exposes the weakness of master planning in difficult policy arenas; Dakar, Senegal (Chaudhuri) suggests that some authorities are not working in harmony in developing UA policy; while the case study of Harare (Mutonodzo) presents a more nuanced – and probably realistic – view of policy makers having divergent opinions on UA. There are, in sum, both positive and negative attributes to the practice. As the primary driver of UA is financial, economic incentives are probably the most effective way to encourage responsible UA. Meanwhile, health risk is the most negative attribute of UA and policy must be more restrictive to mitigate risk. Whatever the case may be, and herein lies the lesson, a balanced approach between encouragement of certain practices and the restriction of others is needed.

The good news is that evidence proves governments are starting to listen. The declarations of Harare and La Paz demonstrate the increased buy-in of decision makers for UA – and such declarations are leading to actions. Since 2000, for example, Peru, Brazil, Ghana and China have created national sectoral policy programmes on UA. This is remarkable for a sector that was hardly recognized a decade ago.

All of these questions feed into the type of policy used by municipal authorities regarding UA. For instance, in the case of poverty reduction, cities must look at specific interventions targeting the poor. To increase food availability, a policy that targets wealthier farmers in larger peri-urban farms may be more important. In fact, the more we learn about UA, the more nuanced policy support becomes.

Climate change

The failure to find consensus on how to address climate change, combined with the increase in resources available for both the mitigation and adaptation work related to climate change, has impacted on discussions around the whole issue of the environment. While energy is at the centre of the debate around climate change, one should consider what the energy arguments are in favour of UA. Certainly less transportation of food from rural areas to cities directly reduces energy needs. Also, the marginal but still-relevant use of rooftops for agricultural production has an important influence on energy conservation. UA increases reliance on local sources of food instead of supporting the ongoing dependency upon the transportation of products from distant places.

Meanwhile, evidence is growing that climate change is reducing the availability of fertile arable land due to desertification. The Millennium Ecosystem Report suggests that an estimated 10–20 per cent of dry land is affected by land degradation (Millennium Ecosystem Assessment, 2005). This change is leading to an increased migration to cities and livelihoods that are put under stress. Moreover, human settlements are often located on the most arable land. Population and cities, however, continue to grow, as does the demand for food. Nor is technology yielding the same improvements in food production efficiency as it once did. In fact, evidence shows that global yields of important crops such as maize, wheat and barley are decreasing with the advent of climate change (Lobell and Field, 2007). The rate of change is outpacing the capacity of technological inputs to compensate.

Changing patterns of consumption are also having a significant influence on the entire agricultural system. As wealth increases, diets change. More meat is being consumed in the rapidly growing economies, particularly in Brazil, Russia, India and China (BRIC countries). Meanwhile, meat production is heavily reliant on grain for feed. Naturally, any change in rural agricultural patterns influences the form of UA in cities. Some patterns are having a serious impact on the scope of food production. For example, rising oil prices and the shift towards fuel crops for biofuel are upsetting food security by increasing the prices of basic foods such as corn and wheat. Clearly, the issue of climate change, transportation and the internationalization of food production and distribution networks must play a role in future research.

Changing our methodological approach: Two lessons

In the book *Agropolis* (Mougeot, 2005), Mougeot pleaded with researchers not to ignore the large amount of grey literature that exists on UA. Grey literature has provided a great deal of baseline information for dozens of city case studies on UA. Much of this is captured on websites such as City Farmer (http://www. cityfarmer.org), and Urban Agriculture Online (http://urbanagriculture. wordpress.com) while still more is captured by the Resource Centres on Urban Agriculture and Food Security's (RUAF; http://www.ruaf.org) *Urban Agriculture Magazine*. These sources act as a repository which are informing more sophisticated debates concerning the global and local importance of UA, while a review of grey literature helps frame the debate on the advantages and disadvantages of UA. This ample resource should be captured by researchers in order to frame new projects. An important contribution to the field would be a comprehensive effort to synthesize existing grey literature.

A second lesson taken from the studies presented in this book is that research on this UA requires a wide variety of methods and an interdisciplinary approach to research. Social and natural sciences are now frequently integrating ethnographic and anthropological analysis into their research protocols. This change is significant because on the one hand an interdisciplinary approach to research yields high value outputs, but on the other hand it requires financial and human resources that are often unavailable. Therefore, the second lesson is that researchers should make the effort to triangulate and corroborate their findings with sources and opinions from outside their field and/or their background. For instance, a natural scientist whose work is reviewed by a sociologist will carry added weight and vice-versa.

What next?

AGROPOLIS was a major investment on the part of IDRC and we are proud of the results. In 2007, IDRC launched a new initiative, 'Ecopolis', which moves to support research on a number of urban environmental issues. These include UA as well as water and sanitation, solid waste management, urban vulnerability and land tenure. Ecopolis will not only emphasize traditional academic research, but will also include resources for designers who are engaged in technical work that enhances higher value UA production. Although UA has made great strides in terms of acceptance, now the greater challenge is to improve architects' and planners' awareness of and sensitivity to including 'green' and agricultural features into the design process. In developing countries, where planning systems can be weak, this means enforcing protective measures that are included in zoning provisions.

The relative importance of UA is on the rise, both as an economic activity and also as a contribution to environmental management. What is absolutely necessary – and the chapters of this book corroborate this conclusion – is for a continued push to legitimize UA, where it is still illegal in the eyes of policy makers and planners, and to draw on the positive experiences of some cities to ensure the viability of the UA economy.

REFERENCES

Agrawal, M., Singh, B., Rajput, M., Marshall, F. and Bell, J. N. B. (2003) 'Effect of air pollution on peri-urban agriculture: a case study', *Environmental Pollution*, vol 126, no 3, pp323–329

Amoah, P., Drechsel, P., Henseler, M. and Abaidoo, R. C. (2007) 'Irrigated urban vegetable production in Ghana: microbiological contamination in farms and markets and associated consumer risk groups', *Journal of Water and Health*, vol 5, no 3, pp455–466

CAB, Abstracts and academic search premier, http://search.ebscohost.com (accessed April 2008)

Dreschel, P., Graefe, S. and Fink, M. (2007) *Rural–Urban Food, Nutrient and Virtual Water Flows in Selected West African Cities*, IWMI Research Report no 115, IWMI, Accra

IWMI (2007) Annual Report 2005–2006, IWMI, Colombo, Sri Lanka

Lobell, D. B. and Field, C. B. (2007) 'Global scale climate–crop yield relationships and the impacts of recent warming', *Environmental Research Letters*, vol 1, no 1

Mara, D. and Feachem R. (2001) 'Taps and toilets for all', in *Water21*, August 2001, IWA Publishing, Colchester, UK

Millennium Ecosystem Assessment (2005) *Millennium Ecosystem Assessment: Synthesis Report*, Island Press, Washington DC

Mougeot, L. J. A. (2005) *Agropolis: The Social, Political and Environmental Dimensions of Urban Agriculture*, Earthscan, London, UK

Index

dietary diversity, 12, 15, 22, 26, 28, 32, 74
and crop diversity, 30
see also crops
diseases, 32, 40–1, 44, 46, 120, 154, 156,
170, 177, 190, 193
crop diseases, 40–41, 44, 46, 156, 170
water-borne, 9, 59, 112, 125
see also pathogens

economics, 3–4, 52–53, 91, 237–238
capital, 60–61
expenditure, daily, 155–156
value of urban agricultural goods, 15
value of urban agriculture undervalued,
18
see also food
education, 82, 97, 155, 176
agriculture, 21, 26, 32, 42, 58–59, 65–6,
97, 100, 188–189
nutrition, 22, 27, 32–33
participatory education and action
research (PEAR), 12, 182,
184–185, 189–194, 197
see also Dakar, case study in
electricity, 83, 94, 151, 154–155, 174, 223
hydro-electric, 125
employment,
fertilizer use reduces, 1
informal, 86, 181
urban agriculture creates, 1, 4, 6–7, 147,
167, 170, 227
see also unemployment
energy adequacy, 74, 78, 84–86
environment, 16–17, 86, 105, 125, 151,
181–182, 189–190, 240–241
political ecology, 10, 203
social ecologies, 8, 13–14, 201–213
sustainability, 88, 100, 106, 149, 163,
202
esteem, *see* self-perception

families, 63
size of, 60, 80–82, 85
urban agriculture supports, 1, 57–58,
61–62, 68
'farm to fork' approaches, 7, 16, 239
fencing, 22, 24, 27, 32, 204
fertilizers, 1, 9–10, 41, 44–46, 59, 86
as pollutants, 156, 158–161
see also compost; manure; poultry; yield

finance, 18, 181
see also banking; credit
flooding, 154, 156, 181, 221–223, 227,
231, 233
floodplains, 221–223, 227, 231, 233
food, 15
choice, 1
distribution, 5, 53, 74, 87, 107, 148, 167,
221, 223, 228, 241
miles, 1, 7, 228
policy, 7–8
prices, 6, 73, 83, 94, 221, 228
see also food security; hygiene
Food and Agriculture Organization
(FAO), 7, 15, 37, 91–92, 236
food security, 1, 4, 7, 13, 15, 22–23,
28–32, 181, 235–236
analysis, 78–79, 82–84
and crop diversity, 22
definitions of, 74, 92
and energy adequacy, 84–86
food entitlements, 15, 92–93, 95
and interventions, 30–31
monitoring, 73
and poverty eradication, 22, 91
status, 79
urban agriculture's contribution to,
studying, 95–100, 235–236
see also Pearson product moment
fruit, 26, 28, 109, 148, 153, 156, 206–211,
213, 223

gardens, 13, 32, 153, 202, 204–205
market gardens, 148–151, 153–156,
158, 160–163
Gaza Strip,
population density, 5
gender, 12, 13–14, 50, 54, 58, 62–66, 67,
170, 203
and farming efficiency, 13, 210,
212–213
female-headed households, 13, 80, 94,
97
and mapping, 205–207
and markets, 185, 210
women and children as major actors in
urban agriculture, 13, 62–63, 67,
100, 171–173, 185, 187–188
and working hours, 13, 63, 173, 204
germination, 38, 43–44

20 years of publishing
for a sustainable future

Sustainable Agriculture and Food Security in an Era of Oil Scarcity

Lessons from Cuba

Julia Wright

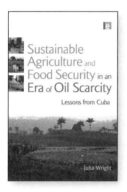

'This is a topical book, now that climate change, the end of cheap oil, growing international disparities and the untenability of the current approaches lead to the insight that business as usual is not in order. The world can learn a great deal from the way Cuba handled its food crisis.' *Niels Roling, Emeritus Professor, Wageningen University, The Netherlands*

'The author has a deep experience of recent transitions in Cuba, and there will be great interest in this book.' *Professor Jules Pretty, MBE, University of Essex, UK*

When other nations are forced to rethink their agricultural and food security strategies in light of the post-peak oil debate, they only have one living example to draw from: that of Cuba in the 1990s. Based on the first and – up till now – only systematic and empirical study to come out of Cuba on this topic, this book examines how the nation successfully headed off its own food crisis after the dissolution of the Soviet Bloc in the early 1990s.

By identifying the challenges faced by Cuban institutions and individuals in de-industrializing their food and farming systems, this book provides crucial learning material for the current fledgling attempts at developing energy descent plans and at mainstreaming more organic food systems in industrialized nations. It also informs international policy on sustainable agriculture and food security for less-industrialized countries.

Dr Julia Wright has worked for over 20 years in sustainable and organic agriculture research and development. She is currently Head of Programmes at Garden Organic (formerly the Henry Doubleday Research Association) in the UK, and on the steering committees of the Bioregional Development Group, Transition Town Leamington, and Canalside Community Foods CSA.

Hardback £60.00 • 256 pages • 978-1-84407-572-0 • November 2008

For more details and a full listing of Earthscan titles visit:

www.earthscan.co.uk

20 years of publishing
for a sustainable future

The Transformation of Agri-Food Systems

Globalization, Supply Chains and Smallholder Farmers

Ellen B. McCullough, Prabhu L. Pingali and Kostas G. Stamoulis

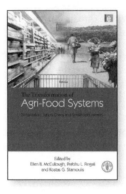

'There should be a good market for this book. The topic is very timely and a major theme of the new World Development Report 2008. The editors and contributors are world class.' *Derek Byerlee, World Bank*

'This is a topic of wide interest and high policy importance. The depth of coverage and excellent synthesis should ensure that the book will have a substantial market in high-level undergraduate and graduate courses in agricultural development. It will have a solid readership among development economists and policy makers as well.' *Mark Rosegrant, International Food Policy Research Institute*

The driving forces of income growth, demographic shifts, globalization and technical change have led to a reorganization of food systems from farm to plate. The characteristics of supply chains – particularly the role of supermarkets – linking farmers have changed, from consumption and retail to wholesale, processing, procurement and production.

This has had a dramatic effect on smallholder farmers, particularly in developing countries. This book presents a comprehensive framework for assessing the impacts of changing agri-food systems on smallholder farmers, recognizing the importance of heterogeneity between developing countries as well as within them.

Published with FAO.

Ellen B. McCullough and **Prabhu L. Pingali** were, at the time of preparation of this book, economists at the Food and Agriculture Organization (FAO) of the United Nations, Rome, Italy. **Kostas G. Stamoulis** is an economist at FAO, Italy. **Prabhu Pingali** is the former president of the International Association of Agricultural Economists.

Paperback £39.95 • 416 pages • 978-1-84407-569-0 • August 2008

For more details and a full listing of Earthscan titles visit:

www.earthscan**.co.uk**

20 years of publishing
for a sustainable future

Convivial Urban Spaces
Creating Effective Public Places

Henry Shaftoe

'Brilliant! This thoughtful and illuminating book fills a significant gap in urban design literature and is a must-have reference book for anyone concerned with the creation or management of the public realm.' *Michele Lavelle, Partner in 4D Landscape Design*

'A book that inspires philosophers to consider and develop Shaftoe's message; and a book rich in good practices noted by a traveller and teacher with the hawk-eye of an architect and photographer.' *Tobias Woldendorp, Senior CPTED Consultant at DSP-groep, Amsterdam.*

Despite developments in urban design during the last few decades, architects, urban planners and designers often continue to produce areas of bland, commercially led urban fabric that deliver the basic functional requirements of shelter, work and leisure but are socially unsustainable and likely generators of future problems.

Convivial Urban Spaces demonstrates that successful urban public spaces are an essential part of a sustainable built environment. Without them we are likely to drift into an increasingly private and polarized society, with all the problems that would imply. Taking a multidisciplinary approach, this book draws on research, and the literature and theory of environmental psychology and urban design, to advance our understanding of what makes effective public spaces.

Practical guidance is illustrated with case studies from the UK, Spain, Germany and Italy. The result is a practical and clearly presented guide to urban public space for planners, architects and students of the urban environment.

Henry Shaftoe is a senior lecturer, specializing in safer, better neighbourhoods, at the University of the West of England. He has also worked as a consultant with the Safe Neighbourhoods Unit (a national not-for-profit organization) and undertakes research into many aspects of urban safety and quality of life.

Hardback £39.95 • 208 pages • 978-1-84407-388-7 • June 2008

For more details and a full listing of Earthscan titles visit:

www.earthscan.co.uk

20 years of publishing
for a sustainable future

The International Journal of
Agricultural Sustainability

Editor in Chief: Jules Pretty OBE, University of Essex, UK

ISSN: 1473-5903
E-ISSN: 1747-762X
Frequency: 4 issues per year.

Published by Earthscan from volume 4 onwards

'Deserves to be made widely available in all institutions
concerned with rural economics and development.'
Times Higher Education Supplement

The International Journal of Agricultural Sustainability (IJAS) is a cross-
disciplinary, peer-reviewed journal dedicated to advancing the understanding
of sustainability in agricultural and food systems.

IJAS publishes both theoretical developments and critical appraisals of
new evidence on what is not sustainable about current or past agricultural
and food systems, as well as on transitions towards agricultural and rural
sustainability at farm, community, regional, national and international levels,
and through food supply chains. It is committed to clear and consistent use
of language and logic, and the use of appropriate evidence to substantiate
empirical statements.

IJAS increases knowledge on what technologies and processes are
contributing to agricultural sustainability, what policies, institutions and
economic structures are preventing or promoting sustainability, and what
relevant lessons should be learned.

For more details and a full listing of Earthscan titles visit:
www.earthscan**.co.uk**

Earthscan e-Alerts
Sign up today!

Keep up to date with Earthscan's new titles in all aspects of sustainable development.

Sign up today to be reminded of new publications, forthcoming events and details of exclusive special offers.

E-alerts also include links for inspection and review copy requests.

Visit **www.earthscan.co.uk** to sign up for our monthly e-newsletter and subject-specific book e-alerts in the following subjects:

- Agriculture, Food and Water
- Architecture and Construction
- Business and Environmental Management
- Cities and Infrastructure
- Climate and Climate Change
- Design
- Development
- Ecology, Biodiversity and Conservation
- Economics
- Energy
- Environmental and Sustainability Assessment
- Forests
- Health and Population
- Natural Resource Management
- Politics, Governance and Law
- Risk, Science and Technology
- Sustainable Development
- Tourism

Once you have registered you can log in using your email address and password, and you can manage your e-alert preferences on your member's page. If you have any queries about your membership or anything else, don't hesitate to email us at **earthinfo@earthscan.co.uk** or give us a call on **+44 (0) 20 7841 1930.**